D1590275

No-Tillage Agriculture
Principles and Practices

No-Tillage Agriculture
Principles and Practices

Edited by

Ronald E. Phillips
Department of Agronomy
University of Kentucky

Shirley H. Phillips
Associate Director for Extension
College of Agriculture,
University of Kentucky

VAN NOSTRAND REINHOLD COMPANY
NEW YORK CINCINNATI TORONTO LONDON MELBOURNE

Library of Congress Catalog Card Number: 83-5827
ISBN: 0-442-27731-8

Manufactured in the United States of America

Published by Van Nostrand Reinhold Company Inc.
135 West 50th Street
New York, New York 10020

Van Nostrand Reinhold Company Limited
Molly Millars Lane
Wokingham, Berkshire RG11 2PY, England

Van Nostrand Reinhold
480 Latrobe Street
Melbourne, Victoria 3000, Australia

Macmillan of Canada
Division of Gage Publishing Limited
164 Commander Boulevard
Agincourt, Ontario M1S 3C7, Canada

15 14 13 12 11 10 9 8 7 6 5 4 3 2 1

Library of Congress Cataloging in Publication Data
Main entry under title:

No-tillage agriculture, principles and practices.

Includes index.
1. No-tillage. I. Phillips, Ronald E. II. Phillips, Shirley H.
S604.N6 1983 631.5′8 83-5827
ISBN 0-442-27731-8

Contributors

Robert L. Blevins, Professor of Agronomy,
University of Kentucky

Wilbur W. Frye, Associate Professor of Agronomy,
University of Kentucky

Ronald E. Phillips, Professor of Agronomy,
University of Kentucky

Shirley H. Phillips, Associate Director for Extension,
College of Agriculture, University of Kentucky

M. Scott Smith, Assistant Professor of Agronomy,
University of Kentucky

Grant W. Thomas, Professor of Agronomy,
University of Kentucky

William W. Witt, Associate Professor of Agronomy,
University of Kentucky

Preface

No-tillage cropping systems and concepts have evolved rapidly since the early 1960s and are attracting attention worldwide. The rapid growth and interest is associated with increasing pressures for food production from a fixed land resource base with degrading effects of erosion, soil compaction and other factors becoming more noticeable.

Research programs have provided many answers and identified new technology needed for success of the no-tillage crop production system in the past two decades and this has resulted in a rapid rate of adoption. Farmers played an important role in the early stages of development of the system and continue to play an important role in its improvement and rapid rate of adoption.

This book provides an inventory and assessment of the principles involved in no-tillage concepts and addresses the application of the technology to practical production schemes. Selected authors and contributors have long been associated either in no-tillage research or application. They represent many disciplines interfacing with the complex interactions of soil, plant and environment. Personal observations by the authors in many geographic sectors of the world indicate the principles to be valid but application of the principles to be less uniform. The application of no-tillage principles requires considerable modification as variations in soil and/or climatic conditions are encountered in different regions of the world.

No-tillage today appears to be relevant to the large commercial producer as well as the smaller farmer producing food and fiber for his family; the principles of no-tillage crop production apply equally well for both. Reduction of energy, whether it be fossil fuel or manpower, will remain an important consideration in the future, with specific implications for the spread of no-tillage to all parts of the world.

It has been estimated that U.S. farmers, in preparing land for row crops, move annually the equivalent amount of soil required to build a superhighway across the continent. The energy cost and soil erosion losses associated with this soil disturbance will be reduced as more hectares of crops are grown by the no-tillage production system. Recent research indicates that legumes can be used as winter cover crops in the no-tillage system and provide significant biological nitrogen to the row crops which follow, resulting in a decrease of the amount of nitrogen fertilizer needed. This will often result in less pollution of ground water by nitrogen. In addition, pesticides applied to soils in no-tillage have been observed to be inactivated more and degrade faster. For this reason higher pesticide rates may be required for adequate control. On the other hand, this and the reduction in surface run-off minimize the possibility of pesticide contamination of the environment.

The start of the twenty-first century may signal the end of a period of 200 years in developing the plow and the initiation of a new system that deletes or at least minimizes the use of this implement in crop production. The deletion of the moldboard plow in no-tillage could have as great an impact on agriculture in the future as the moldboard plow has had in the past. However, continued research and application of new technology in no-tillage will be required in order that this happen. Future demands on land resources require that modifications in production practices be made in order to supply the ever-increasing demand of food and fiber. No-tillage will doubtless continue to be one of the most important of these production practices, one that protects the soil, conserves water and reduces energy consumption.

Shirley H. Phillips
Ronald E. Phillips
Lexington, Kentucky

Contents

No-Tillage Agriculture
Principles and Practices

1
Introduction

Shirley H. Phillips
Associate Director for Extension
College of Agriculture, University of Kentucky

No-tillage may be defined as the introduction of seed into unplowed soil in narrow slots, trenches or bands of sufficient width and depth for seed coverage and soil contact. Thus a new and exciting concept emerges, eliminating the usual traditional plowing and discing operations for ordering of soil in crop production.

Many terms are used in reference to no-tillage. The terminology no-tillage, zero till, no-till, zero tillage, mulch tillage and others commonly used do not represent the concept, as some tillage action occurs in the incorporation of seed. Eco system, stubble planting, mulch planting and chemical fallow is used in identifying similar systems of small grain production.

No-tillage probably is indicative of the complete cycle in a 100–200 year agrarian system. The American Indian produced food crops by disturbing the soil surface with crude hand tools to a sufficient degree needed to cover the seed, and the animal products, such as fish, placed near the seed provided plant nutrients. Similar systems developed in other parts of the world, involving burning or slashing of existing vegetation and creation of residues prior to planting seed.

At this point in the history of the world, power for tillage was limited and vast land resources were available which dictated the crop production system in use at that time.

Domestication of animals and substitution of animal power for human energy increased man's ability to till land and produce food. Centuries elapsed between the use of crude, wooden tillage implements and Thomas Jefferson's development of a mathematical formula for the mold-board plow. In 1796 Charles Newbold received a patent for a cast iron plow. In the early 1830s the steel moldboard plow was introduced. Continued refinement of primary and secondary tillage implements evolved into highly sophisticated, large units into the 1980s, which move each year the equivalent amount of soil that would be required to build a superhighway from New York to California.

The introduction of steam to power tractors in 1868, followed by internal combustion engines for tractors in the early 1900s, triggered the horsepower era that allowed the development of huge 300–400 horsepowered tractors in use today with the capabilities of pulling enormous tillage implements.

Attention turned to reduced tillage in the late 1940s with the introduction of plant growth regulators developed during World War II. Klingman, in North Carolina in the late 1940s reported no-tillage work. In 1951, K. C. Barrons, J. H. Davidson and C. D. Fitzgerald of the Dow Chemical Co., reported successful application of no-tillage techniques. In the 1960s M. A. Sprague, in New Jersey, reported pasture renovation using chemicals as a substitute for tillage. L. A. Porter, of the Levin Horticulture Research Station in New Zealand, in the early 1960s reported on strawberry production without tillage, followed by A. E. M. Hood, and R. S. L. Jeater at Jealott's Hill, England, with small grain. In 1968 W. Scharbau, Deutschland, West Germany, reported on rape and small grain no-tillage. Hence, worldwide interest in no-tillage research and application was well underway.

Refinement of the system in the U.S. can be traced to George McKibben, Illinois; D. B. Shear and W. W. Moschler, Virginia; G. B. Triplett and D. M. VanDoren, Ohio; J. F. Freeman, Kentucky; W. M. Lewis and A. D. Worsham, North Carolina; and many others in the Midwest and Southeast. Speight, Reynolds Tobacco Co., North Carolina, reported on double cropping with winter small grains and soybeans using the no-tillage approach which signaled double cropping opportunities as the potentially most significant contribution from the total no-tillage introductory concept.

As a parallel to the development of plant growth regulators during World War II, machines capable of deep fast tillage were being developed. In retrospect it is apparent the technology was emerging to a point where no-tillage would ultimately come into focus. Few would envision distinctly the different thrust and philosophy merging into the system that has gained worldwide acceptance today.

Several practices became widely discussed in the 1950s, yet sparingly used, including wheel track planting and plow plant. Perhaps these developments and considerations were needed in a systematic continuance of reduced tillage to make present systems acceptable and technically sound for producers.

Wheel track planting was a method where soil was turned or plowed, followed by planting with the planter aligned and situated to plant directly behind and in the depressions created by tractor tire track. The tire action firmed the soil and eliminated the need for discing and other soil preparation. At the same time, one U.S. farm equipment manufacturer was developing a "plow-plant" system. This system, like wheel track planting, involved plowing of the land and planting in the same operation without additional seedbed preparation. Farmers did not accept these systems because of a major timing conflict resulting from all plowing being delayed until planting. Weather risk and peak labor requirements negated the benefits gained by these systems.

Conventional tillage could be generally described at this time to include fall or spring plowing with two to four discings prior to planting. Post emergence cultivations were standard until the late 1940s to control weeds.

Minimum tillage proponents advocated reduced amounts of tillage in the late 1940s and started the use of selective plant growth regulators for post emergence weed control. This system is the major cultural system being used in clean, cultivated crops throughout the world today. These systems tend to follow plowing with one discing, and preemergence herbicides used in lieu of additional discings and cultivation, and make up the largest percentage in the minimum tillage category in the world.

REASONS FOR TILLAGE

Inquiries as to reasons farmers plow usually center around controlling weeds more than all others. Many operators would add that plowing is needed to cover surface residues and aesthetic value provided by emerging seedlings in a totally clean soil surface environment. Aeration of the soil was considered important and possibly needed when the structure was altered by overzealous soil movement and compaction created by heavy equipment making many passes over the field.

Pest control achieved through covering of crop residues was considered essential in short crop rotations or sequences and is cited as a reason for tilling soils between crops by many growers and pathologists. Leveling or shaping of field by plowing and discing was needed after cultivation or from harvesting operations during periods of heavy rainfall creating unfavorable conditions, and is given as a reason for tillage. Incorporation of fertilizers and other soil amendments were important reasons under conventional or minimum tillage operations. Soil temperature could be increased by plowing, covering residues and increasing solar heat absorption in temperate regions with shorter or minimum growing season for crops. Stimulation of root growth through tillage was a conception commonly expressed and had a seemingly logical base of support. The pressure of tradition dictated farmers to continue routines established over years of successful production by using implements and practices accepted and unchallenged by either the scientific or the farm community.

These reasons had validity when one considers alternatives or options available at this point in time. The production demands could be met using these proven methods until the need for food and fiber increase associated with World War II placed great stress on world soil resources. Industrialization, population explosions and economic changes continued to increase demands on land resources with predictable further expansion and demands.

Factors that Contributed to the Increased Interest in No-Tillage Systems

It is conceivable that questions will be raised relative to an early cultural system that was used extensively, passed, lay dormant for many years, then reappeared on the horizon and captured interest over the world some 200 years later. There were many factors that came into focus simultaneously that were responsible for this resurgence and created an atmosphere for a system such as this to gain renewed attention.

The development of plant growth regulators and, in particular, the non-selective contact weed control materials, that became commercially available provided the ability to eliminate or severely suppress existing vegetation. These materials provided a base for recommendations and allowed for consistent yields and predictable performance.

Population shifts to urban areas depleted farm labor pools in many parts of the world. Capital was invested in substitutes such as pesticides, equipment and other production items. The conventional tillage methods of crop production were most labor-intensive. This compelling force would have great impact on the level of interest and momentum in shifting to less tillage techniques in many of the developed countries. It should be emphasized at this point that in countries with surplus farm labor situations, such as third world countries, the no-tillage system can aggravate this problem and must be included in evaluation of adoption of these crop production systems in these areas.

Farm equipment manufacturers introduced modified no-tillage type planting equipment with capabilities to place seed into uncultivated, unprepared soils. Prior to this period farmers were constructing and modifying existing equipment with inconsistent performance and results.

The introduction of no-tillage type equipment was met with understandable resistance by the industry, which was moving to more powerful tractors and larger tillage implements. Less tillage concepts eliminated or at least threatened the need for this type of equipment. Added longevity and the elimination of plows, discs, cultivators and other tillage tools was a potential threat for future sales. In the final analysis, equipment was available and did a creditable job in meeting the changing environment created by moving from prepared seedbeds to no-tillage concepts.

Planting delays created by unusual heavy rainfall occurred over many parts of the world in the late 1960s. Farmers faced with normal and associated reduced yields by delayed planting were more receptive to practices that would save time in establishing crops such as was afforded by the adoption of no-tillage. These factors merging within the same time frame made no-tillage concepts more palatable and the adoption rate was accelerated. The absence of either the planting equipment, pesticides or reduced farm labor, or climatic pressure, would have retarded the adoption of no-tillage on farms.

Advantages of No-Tillage

All crop cultural systems that create interest to the extent demonstrated by no-tillage must have demonstrated apparent advantages over traditional methods. Several characteristics inherent to this system can be enumerated.

Erosion Control. The pattern of adoption by areas of the world reflect the importance of this advantage. Much of the world's soils devoted to crop production has limitations and restrictions associated with wind and water erosion.

Many studies confirm that no-tillage agriculture reduces soil erosion to almost zero. McGregor *et al.* (1975) found that on a highly erodible soil in Mississippi, erosion was reduced from 17.5 mt/ha to about 1.8 mt/ha when the no-tillage system was used, and Triplett and VanDoren (1977) found that no-tillage reduced soil erosion by as much as 50-fold.

Reduced Fuel. Fuel savings result from land preparation operations, with harvest, processing and transportation remaining about equal with the various cultural systems in use. Peak demands for fuel occur during land preparation and plantings. This period normally follows heavy demands for home heating in temperate zones and potential depleted reserves. No-tillage can alleviate to an extent these competitive requirements. Diesel fuel requirements for land preparation is about 60–75 percent less for no-tillage than for conventional tillage. Additional energy will be required for contact herbicides used in no-tillage; however, a net reduction in energy will result from no-tillage, as discussed in Chapter 6.

Flexibility in Planting and Harvest. Infrequently climatic conditions retard land preparation and interrupt normal planting periods. No-tillage offers the opportunity to plant without waiting for sufficient drying time for tillage. Untilled soil provides improved trafficability in planting, pesticide application and harvesting.

Increased Land Use. The Soil Conservation Service has recommended increasing the intensity of land use within capability classes when no-tillage systems are in use. Classes IIe, IIIe and IVe may be upgraded one class with no-tillage and incorporation of other conservation practices. The impact on each individual farm's capacity to produce food and fiber could reach magnitudes of 10–30 percent within acceptable conservation standards on land subject to erosion losses. Moisture retention and reduced evaporation can expand production into other soils considered marginal with conventional tillage methods.

Reduced Labor Requirements. Comparisons of no-tillage and conventional tillage labor requirements were made based on records of farmer participants in the Kentucky Farm Business Analysis, which indicate a 50 percent labor reduction in land preparation and planting in favor of no-tillage. Typical acreages of no-tillage crops grown by farmers are 2-320 hectares per man as compared to 0.8-160 hectares per man for conventional tillage on mechanized farms.

Improved Water Retention, Reduced Evaporation and Other Crop-Soil Relationships. Blevins *et al.* (1971) conducted studies to determine soil moisture under killed sod and conventional tillage using corn as the growing crop. These studies indicated 19 percent higher moisture in the no-tillage plots with the higher amounts in the 0-15-cm depth. This extra amount of plant-available water was sufficient to carry the growing crop for short drought periods of two to three weeks. Increased infiltration and reduced runoff is another factor which, coupled with less evaporation, presents a more favorable environment for increased yields. Observations over 5-10 years suggest less organic matter decrease, improved soil structure on most soil types, and substantial soil compaction.

Wiese and Unger (1974) in Texas report efficiency in irrigation using the no-tillage system for wheat and grain sorghum production: 6.5-13 cm of irrigation water per hectare can be conserved in comparison to complete tillage. This will become more important as energy costs for irrigation increase and available irrigation water supplies are diminished.

The no-tillage system is adaptable to farms regardless of size. The no-tillage concepts are being used by farmers using hand labor in various parts of the world. Other growers produce crops on 1-20 hectares using the same principles as in a highly mechanized system. Livestock producers can grow forages and silage crops along with producers of grain and fiber for the cash market. Multi-cropping acreage discussed in Chapter 11 has increased many fold. The adoption of no-tillage has allowed new crop systems and sequences to develop throughout the entire world.

No-Tillage Techniques are Adaptable to Most Crops. A system that started basically by growing corn in a killed sod has permeated the agricultural sector to offer opportunity to most all crops. Root crops may have less potential than those crops with harvested products produced above the soil surface.

Equipment Requirements are Lower. The basic implements required for no-tillage include planter, pesticide applicator and harvest equipment. Tractor horsepower requirements are comparatively low and expected life extended by decreased hours of operation. Multi-cropping can spread machinery cost over more hectares without adding to land inventory.

Disadvantages of No-Tillage

As with most systems, disadvantages should be considered by potential producers. No-tillage is not an exception. The system appears to be rather simple with casual observation. Unfortunately, this is not true; a higher level of management or at least new technology will be required by growers. Fertilizer usage and timing may vary from that of conventional tillage. Producers must learn planting techniques that will provide sound planting principles, depth, seed-to-soil contact and seed cover in the absence of loose pulverized soil delivered by tillage.

Soil Temperature. Soil temperature regimes, especially in the surface layers, are strikingly different, which can create problems in temperate zones. Temperatures can run 2–10°C lower and aggravate cool soil situations for warm season crops. The reverse will be true in the tropics, where soil temperatures are too high for optimum plant growth and development. What is a disadvantage in temperate regions can be an advantage in tropical regions. Late frost and light freezes will be more damaging on no-tillage crops than on tilled soils.

Weed Control. Weed control will be a greater problem without plowing, and shifts in weed populations will be more pronounced. Technology of application, selection of herbicides, crop rotations and other information must be adopted by the grower using the no-tillage concept.

Insect, Rodent and Disease Incidence. Plant pests tend to be of more consequence in untilled soils with crop residues at the surface. The levels of infestations have not been as much of a problem as anticipated with the micro environment created in an untilled soil.

Aesthetics Associated with No-Tillage. It has been difficult for many farmers to accept the ragged appearance of no-tillage fields. Crop residues at the surface with varying degrees of dying vegetation represent a drastic departure from clean soil surface, and finely pulverized and sized soil particles associated with conventional tillage. This uneven, unkept, untidy appearance continues until the crop canopy is formed. Ridicule of non-practitioners is paramount during the early stages of establishment and has had a social impact on adoption rate.

PRESENT PHILOSOPHY IN NO-TILLAGE

There appear to be diverse philosophies developing in the introduction and expansion of no-tillage. It is fortunate that these exist and they can do much for the future of no-tillage. One trend is to consider the use of no-tillage on the more productive soils and to maximize yield through no-tillage manipulation. This trend centers in the Americas. Another direction is to take marginal soils with high-risk situations and modify no-tillage to enhance yields or develop the ability to use these soils in cultivated crops. Much of the positive direction for this concept centers in Europe. It would be unwise to assume that the scientific base that exists today in no-tillage will be adequate for the future. The history of modern no-tillage technology is short, in the magnitude of less than 20–40 years. Efforts in developing information have been modest but are increasing in all sectors of the world. The interest on the part of researchers, combined with economic factors and an innovative group of farmers, will generate systems of crop production for future production of dependable supplies of food and fiber for the world's consumers.

REFERENCES

Blevins, R. L., D. Cook, S. H. Phillips, and R. E. Phillips. 1971. Influence of no-tillage on soil moisture. *Agron. J.* **63**:593–596.

McGregor, K. C., J. D. Greer, and G. E. Gurley. 1975. Erosion control with no-till cropping practices. *Am. Soc. Agric. Engr. Trans.* **18**:918–920.

Triplett, G. B., Jr. and D. M. VanDoren, Jr. 1977. Agriculture without tillage. *Sci. Am.* **236**:28–33.

Wiese, A. F. and P. Unger. 1974. Soil water storage and use. *Proc. of a Symposium on Limited and No-Tillage Crop Production.* USDA Southwestern Great Plains Res. Center, pp. 17–26.

2
Effects of Climate on Performance of No-Tillage

Ronald E. Phillips
Professor of Agronomy
University of Kentucky

PRECIPITATION AND EVAPOTRANSPIRATION

Distribution and amount of rainfall and temperature have more direct effects upon performance of no-tillage row crops in dryland agriculture than any other climatic factors. As will be pointed out in Chapter 4, the distribution of rainfall during the growing season is very important to the maximum conservation of soil water due to no-tillage or a mulch. The more rainfall events which occur during the growing season, the greater the difference of soil water evaporation between no-tillage or a mulch and conventional tillage; therefore, the more soil water will be conserved from the no-tillage than from conventional tillage which is then available for transpiration through plants. In central Kentucky the ideal distribution and amount of rainfall to obtain near maximum conservation is about 2.5–3.0 cm of rain per week, a month prior to planting until the crop reaches full canopy, about mid-July (see Table 4-1). From Table 4-1, it can be seen that the average water use, soil water evaporation plus transpiration, by corn for the 1970, 1971, and 1972 and 1973 growing seasons (May through September) on the conventional tillage treatment were 35.5, 51.4, 35.5 and 50.6 cm, respectively. The grain yields were 6,136, 10,841, 10,841 and 8,795 for 1970, 1971, 1972 and 1973, respectively. The water use in 1970 and 1972 was the same, 35.5 cm, while grain yields were drastically different, 6,136 and 10,841 kg/ha. The rainfall distribution was very even in 1972 while in 1970 little rainfall occurred during July and August. The corn in 1970 was water-stressed most of the time during July and August during silking, tasseling and the grain filling period; in 1972 only mild water-stress occurred. In 1971 and 1973, the water use during the growing season was very similar, about 50 cm. The grain yield in 1971 was over 2,000 kg/ha greater than in 1973. Little rainfall occurred in either year in August but the total rainfall was 60.4 and 48.1 cm for 1971 and 1973, respectively. These data point out the importance of distribution and total rainfall to grain yield of corn under non-irrigated agriculture.

Table 2-1 shows monthly precipitation and potential evapotranspiration for April throughout September from 28 selected locations in the U.S. These include locations from both major and minor corn producing states. Potential evapotranspiration exceeded precipitation at 6, 8, 20, 27, 24 and 24 of the 28 locations during April, May, June, July, August and September, respectively. In general, water stress began to develop in June and intensified during July, August and September. At most of these locations, soil water storage of 10–20 cm of plant extractable water in the rooting depth was great enough for evapotranspiration to remain at potential for several weeks during the time when evapotranspiration exceeded rainfall. Row crops at almost all of these locations are water-stressed to a varying degree at least once or more in most years. Dubuque, Iowa and Madison, Wisconsin should have fewer and shorter periods of plant water-stress than the other locations in the corn belt based upon the ratio of precipitation to potential evapotranspiration during the corn growing season (Table 2-1). Corn cannot be grown at Denver, Colorado and Lubbock, Texas due to a lack of water except under irrigation.

Normal monthly precipitation and potential evapotranspiration of selected locations around the world are given in Table 2-2. No-tillage of row crops are practiced in many of these countries. In the northern hemisphere, most of the corn is planted from March to May depending upon the temperature and/or rainfall patterns. In the southern hemisphere, corn planting dates are approximately six months later. The rainfall distribution and potential evapotranspiration for the months of October through March at Buenos Aires, Argentina (the corn growing season in Argentina) are very similar to that of the U.S. corn belt for the months of April through September (the corn growing season in the U.S. corn belt). Corrientes, Argentina, a region where no-tillage of row crops is practiced, has a ratio of precipitation

Table 2-1. Precipitation (P) and potential evapotranspiration (PET), in centimeters, for months of April through September for selected locations in the U.S. (From Denny, 1978 and J. Papadakis, *Climatic Tables of the World,* Buenos Aires, 1961, respectively.)

	APRIL		MAY		JUNE		JULY		AUGUST		SEPTEMBER				
LOCATION	P	PET	P	PET	P	PET	P	PET	P	PET	P	PET	P FOR PERIOD	PET FOR PERIOD	RATIO: P TO PET
Wilmington, Dela.	8.1	7	8.5	9	8.2	12	7.2	13	10.1	12	8.6	10	50.8	63	0.81
Denver, Col.	4.9	7	6.7	10	1.1	16	4.5	18	3.3	17	2.9	14	23.4	82	0.29
Hartford, Conn.	9.5	5	8.9	7	9.0	8	8.7	9	10.0	8	9.0	7	55.1	44	1.25
Athens, Ga.	11.1	10	10.2	12	10.9	15	13.3	15	9.0	15	9.4	13	63.9	80	0.80
Chicago, Ill.	8.6	4	8.6	6	10.6	8	8.8	10	7.0	9	7.7	8	51.3	45	0.14
Peoria, Ill.	11.1	5	9.8	8	10.0	12	9.5	14	7.8	13	9.0	11	57.2	63	0.91
Evansville, Ind.	10.3	7	11.1	10	9.0	13	9.6	14	7.4	14	7.1	12	54.5	60	0.78
Indianapolis, Ind.	9.8	6	10.4	8	10.6	11	9.3	13	7.1	12	7.3	10	54.5	60	0.91
Des Moines, Iowa	7.5	6	10.7	8	12.5	10	8.3	13	8.4	12	7.8	10	55.2	59	0.94
Dubuque, Iowa	10.5	5	11.9	7	13.5	9	10.9	11	10.2	10	11.8	7	68.8	49	1.40
Wichita, Kans.	7.5	8	9.2	9	11.4	13	11.1	17	9.8	18	9.4	13	58.4	78	0.75
Lexington, Ky.	9.8	7	10.6	10	10.9	10	12.3	12	8.6	12	6.8	11	59.0	62	0.95
Columbia, Mo.	9.7	7	11.9	9	11.7	12	9.9	15	8.1	14	11.1	11	62.4	68	0.92
Lansing, Mich.	7.3	5	8.4	7	8.8	9	7.1	11	7.1	10	6.7	8	45.4	50	0.91
Baltimore, Md.	7.8	6	9.1	9	9.5	11	10.3	12	10.7	11	7.9	9	55.3	58	0.95
Minneapolis, Minn.	5.2	5	8.5	7	10.0	9	9.4	12	7.8	10	7.0	8	47.9	51	0.94
Lincoln, Nebr.	6.6	3	9.8	8	13.1	12	8.2	15	9.2	14	8.2	11	55.1	63	0.87
Raleigh, N.C.	7.8	9	8.5	9	9.3	14	12.9	14	12.5	12	9.6	11	60.6	69	0.88
Buffalo, N.Y.	8.0	4	7.5	7	5.6	9	7.5	10	9.0	10	8.3	8	45.9	48	0.96
Spartanburg, S.C.	10.9	9	7.5	10	10.3	12	10.6	12	10.3	12	9.6	11	59.2	66	0.96
Akron, Ohio	8.4	5	10.8	7	8.9	10	9.6	11	7.0	10	6.6	9	51.3	52	0.99
Columbus, Ohio	9.5	6	10.4	8	10.5	11	10.7	12	7.3	12	6.1	10	54.5	59	0.92
Harrisburg, Pa.	7.6	6	9.5	9	7.9	11	9.3	12	8.1	11	6.8	9	49.2	58	0.85
Nashville, Tenn.	10.4	8	10.4	10	8.6	12	9.8	13	8.3	13	7.9	11	55.4	67	0.83
Lubbock, Texas	2.7	10	8.0	13	7.1	17	5.7	18	4.8	18	5.6	14	33.9	90	0.38
Richmond, Va.	7.0	8	8.7	10	9.4	11	7.3	13	12.9	11	9.1	10	54.4	63	0.86
Charleston, W. Va.	8.5	9	8.8	11	8.4	13	12.8	14	9.4	13	7.5	14	55.4	74	0.75
Madison, Wis.	6.8	4	8.7	6	11.0	8	9.6	10	7.7	9	8.5	7	52.3	44	1.19

Table 2-2. Normal monthly precipitation (P) and potential evapotranspiration (PET), in centimeters, around the world. (From Todd, 1970 and J. Papadakis, *Climatic Tables of the World,* 1961, respectively.)

MONTH	LOCATION													
	MOSCOW, U.S.S.R.		MUNICH, GERMANY		LUXEMBOURG, LUXEMBOURG		LONDON, ENGLAND		BARCELONA, SPAIN		PORT ELIZABETH, S. AFRICA		KADUNA, NIGERIA	
	P	PET	P	PET	P	PET	P	PET[1]	P	PET	P	PET	P	PET
January	3.8	1	4.3	1	5.8	1	5.1	2	3.0	4	3.0	8	<0.1	25
February	3.6	1	3.6	1	5.1	1	3.8	2	5.3	4	3.3	7	0.3	25
March	2.8	1	4.8	2	4.8	3	3.6	3	4.8	5	4.8	7	1.3	26
April	4.8	3	6.9	4	5.3	5	4.6	4	4.6	5	4.6	6	6.4	22
May	5.6	6	9.4	5	6.1	6	4.6	5	4.6	6	6.1	7	15.0	16
June	7.4	8	11.7	7	6.4	7	4.1	6	3.3	8	4.6	7	18.0	11
July	7.6	9	11.9	7	7.1	8	5.1	7	3.0	9	4.8	6	21.6	8
August	7.4	7	10.7	7	6.6	8	5.6	7	4.3	9	5.1	6	30.2	7
September	4.8	4	8.1	5	6.1	5	4.6	6	6.6	8	5.8	6	26.9	11
October	6.8	2	5.6	3	6.9	3	5.8	4	8.6	6	5.6	6	7.4	14
November	4.3	1	4.8	2	6.9	1	6.4	2	6.9	4	5.6	6	0.3	20
December	4.1	1	4.8	1	7.1	1	5.1	2	4.6	4	4.3	7	0.1	22
Total	63.0	44	86.6	45	74.2	49	58.2	50	59.6	72	57.6	79	126.8	205
Ratio: P÷PET	1.4		1.9		1.5		1.2		0.8		0.7		0.6	

[1] PET values for Oxford, England.

Table 2-2. Normal monthly precipitation (P) and potential evapotranspiration (PET), in centimeters, around the world. (From Todd, 1970 and J. Papadakis, *Climatic Tables of the World*, 1961, respectively.) (continued)

MONTH	LOCATION													
	BANGALORE, INDIA		BANGKOK, THAILAND		CHUNGKING, CHINA		OSAKA, JAPAN		MANILA, PHILIPPINES		ADELAIDE, AUSTRALIA		SYDNEY AUSTRALIA	
	P	PET	P	PET	P	PET	P	PET	P	PET	P	PET	P	PET
January	0.5	13	0.5	12	1.8	1	4.3	3	2.3	10	2.0	16	8.9	8
February	0.8	16	2.8	13	2.0	3	5.8	3	1.3	12	1.8	16	10.2	8
March	1.0	20	2.8	13	3.8	5	9.7	5	1.8	14	2.5	14	12.7	7
April	4.1	20	5.8	14	9.7	6	13.2	6	3.3	15	4.6	9	13.5	6
May	10.7	17	13.2	12	14.5	7	12.4	8	13.0	13	6.9	6	12.7	5
June	7.4	11	15.2	10	18.0	8	18.8	9	25.4	11	7.6	4	11.7	5
July	9.9	10	17.5	10	14.2	12	15.0	10	43.2	9	6.6	4	11.7	5
August	12.7	9	23.4	10	11.9	14	11.2	13	42.2	8	6.6	5	7.6	6
September	17.0	9	35.6	9	14.7	7	17.8	10	35.6	9	5.3	7	7.4	7
October	15.0	10	25.1	8	10.9	4	13.0	7	19.3	9	4.3	9	7.1	7
November	6.9	10	4.6	10	4.8	2	7.6	5	14.5	9	2.8	12	7.4	7
December	1.0	11	0.3	12	2.0	3	4.8	4	6.6	10	2.5	15	7.4	7
Total	86.9	156	146.8	133	108.2	72	133.6	83	208.5	129	53.5	117	118.3	78
Ratio: P÷PET	0.6		1.1		1.5		1.6		1.6		0.5		1.5	

Table 2-2. Normal monthly precipitation (P) and potential evapotranspiration (PET), in centimeters, around the world. (From Todd, 1970 and J. Papadakis, *Climatic Tables of the World*, 1961, respectively.) (continued)

MONTH	LOCATION											
	RIO DE JANEIRO, BRAZIL		BUENOS AIRES, ARGENTINA		CORRIENTES, ARGENTINA		GUATEMALA CITY, GUATEMALA		MEXICO CITY, MEXICO		OTTAWA, CANADA	
	P	PET	P	PET	P	PET	P	PET	P	PET	P	PET
January	12.4	8	7.9	12	11.9	16	0.8	8	0.5	8	7.4	1
February	12.2	9	7.1	11	11.4	15	0.3	9	0.8	10	5.6	1
March	13.0	8	10.1	9	13.5	13	1.3	9	1.3	12	7.1	2
April	10.7	7	8.9	6	14.2	9	3.0	10	1.8	12	6.9	4
May	7.9	6	7.6	5	8.4	7	15.2	10	4.8	11	6.4	6
June	5.3	7	6.0	4	4.8	6	27.4	11	10.4	9	8.9	9
July	4.1	7	5.6	3	4.3	6	20.3	11	11.4	7	8.6	11
August	4.3	7	6.1	4	3.8	8	19.8	11	10.9	7	6.6	9
September	6.6	7	7.9	5	7.1	10	23.1	11	10.5	7	8.1	6
October	7.9	7	8.6	6	11.9	11	17.3	10	4.1	7	7.4	4
November	10.4	7	8.4	8	13.2	14	2.3	7	1.3	7	7.6	2
December	13.7	8	9.9	11	13.2	16	0.8	8	0.8	8	6.6	1
Total	108.5	88	94.9	84	117.7	131	131.6	115	58.6	105	87.2	56
Ratio: P÷PET	1.2		1.1		0.9		1.1		0.6		1.6	

to potential evapotranspiration similar to Kentucky, a region where no-tillage of row crops has been successful and where about 26 percent of corn and soybeans were grown by the no-tillage system in 1981.

In general, the mulch associated with no-tillage as compared to conventional tillage will conserve the most soil water in climates where evapotranspiration is about 0.5 to 1 cm/week more than weekly rainfall from about a month prior to planting to the time the crop attains full canopy.

SOIL TEMPERATURE

Willis *et al.* (1957) reported that the average soil temperature at the 10-cm soil depth from June 8 to August 16 was decreased by about 0.1–1.5°C due to 5.65 mt/ha of straw mulch as compared to bare soil. They concluded that the average maximum and minimum temperature of the mulched soil during this time period was 24°C and 20°C, respectively; the average maximum and minimum soil temperature for the bare soil was 25.6°C and 20.6°C, respectively. They concluded that the most favorable soil temperature at the 10-cm soil depth for corn growth in Iowa appeared to be about 24°C.

Van Wijk *et al.* (1959) conducted experiments in Iowa, Ohio, Minnesota and South Carolina to test the hypothesis that early season corn growth is decreased by low soil temperatures caused by a mulch of crop residue. The ratio of corn growth on mulched treatments to corn growth on unmulched treatments for 28–54 days after planting was about 0.5, 0.6, 0.9 and 0.95 for Iowa, Minnesota, Ohio and South Carolina, respectively. Both the mulched and unmulched treatments were plowed and a seedbed prepared; straw was then spread uniformly over the soil surface. Similar results would be expected for no-tillage versus conventional tillage. In general, however, under periods of soil water-stress in early season, increased soil water for mulch treatments could offset some of the adverse effects of lower soil temperatures due to the mulch. The same can also be said for no-tillage versus conventional tillage. Van Wijk *et al.* (1959) did not measure soil water or report rainfall for the above experiments.

Burrows and Larson (1962) applied 0, 2.3, 4.5, 9.1 and 18.1 mt/ha of chopped corn stalks distributed evenly over the soil surface in Iowa to evaluate soil temperature at the 0.6, 5.0 and 10 cm soil depth and its influence on growth of corn. The average soil temperatures for a full day, 50 days after planting corn was 4.4°C, 5.0°C and 3.9°C, higher at the 0.6-, 5- and 10-cm soil depths, respectively, for the 0-mt/ha than the 18.1-mt/ha rate. They found that as the amount of mulch increased, corn growth in terms of plant height and dry matter production decreased progressively. As little as 2.3 mt/ha materially reduced the growth of corn early in the season. Dry matter yield was 90 gm/10 plants for the 0 rate of mulch which was decreased to about 15 gm/10 plants for the 18.1 mt/ha rate. They did not report grain yields.

Moody *et al.* (1963) in Virginia compared soil temperature and corn grain yields on treatments where 6.8 mt/ha of straw was used as a mulch over a two-year period. They found at the 10-cm depth, the mean maximum soil temperature difference range was 1–3°C for May and 3.2–3.8°C for June, with the bare treatment (straw plowed under) having the higher temperature. Corn yields over a three-year period averaged 4,152 and 7,016 kg/ha for the bare and mulched treatments, respectively. The average difference in soil water in the 0–46 cm soil depth was 1.2, 2.9, and 2.2 cm for 1958, 1959 and 1960, respectively, in favor of the mulched treatments.

As is typical in Virginia and Kentucky, corn growth in mulched treatments during the early season is depressed compared to bare treatments. Researchers in Virginia and Kentucky have thought that this growth depression is associated with the lower soil temperatures due to mulch. After four to six weeks, corn growth on mulched treatments will surpass growth on bare treatments, and final grain yields are often higher. This is thought to be associated with the higher soil water content of the mulched treatments. It is extremely difficult to separate temperature and soil water effects in the field. The true effect of increased soil water due to mulch or no-tillage corn is often greater than it appears to be from grain yields and/or dry matter production at maturity because of the early depression of corn growth associated with lower soil temperatures.

Soil temperature in no-till corn (killed orchardgrass sod), averaged 10°C lower at the 2.5-cm soil depth as compared to conventional tillage throughout the growing season in West Virginia (Bennett *et al.*,

1973). The average silage yields over a two-year period were 58,468 and 38,192 kg/ha for the no-tillage (killed sod) and conventional tillage treatments, respectively. They also had a no-tillage treatment where the orchardgrass was not completely killed. Silage and grain yields on this no-tillage treatment were equal to or greater than yields of the conventional tillage treatment. In Nigeria, Lal (1974b) reported 9.8°C and 4.2°C difference in soil temperature at the 5 and 20 cm depths, respectively, on sunny days. These differences were about 3–4°C at the 5-cm soil depth and 1–2°C at the 20-cm soil depth on relatively cloudy days. Lal (1974a) also reported soil temperatures at the 10-cm depth as high as 38°C early in the growing season for unmulched soil, whereas on mulched soil the maximum temperature at the 10-cm depth did not exceed 34°C. In Nigeria, where the soil temperature of conventional tillage corn is often above optimum for corn growth, lower soil temperature of no-tillage treatments tends to increase the rate of corn growth rather than decrease it as is the case in temperate climates.

Walker (1970) conducted an experiment to determine the effects of alternating versus constant soil temperatures on corn seedling growth. Corn seedlings were grown at alternating ±3°C about 20°C and constant soil temperature of 23°C and 26°C; air temperature was maintained at 25°C with 12-hour light and 12-hour dark periods. Seedling growth at the alternating temperature was greater than growth at either 23°C or 26°C constant soil temperature.

Willis and Amemiya (1973) reviewed the literature on the effects of tillage management on soil temperature. They concluded that at the 10-cm soil depth, the optimum average soil temperature for corn seedling growth was 24°C. If the average soil temperature at 10 cm was as high as 24°C, then growth and yield were not materially influenced by mulching. Soil temperature affects the physiology of corn through germination, rate of emergence, nutrient uptake, yield, plant height and dry matter production.

The presence of a mulch on the soil surface influences soil temperature in several ways. The presence of mulch reduces the quantity of direct solar radiation reaching the soil surface. It also reflects more of the radiation back to the atmosphere (a mulch usually has a higher albedo than bare soil). Since mulched soil loses less water via evaporation (higher soil water content), the thermal conductivity is greater conducting the heat to a greater depth, thus warming the seed zone less. Also, the heat capacity of wet soil is greater than for dry soil, requiring more heat to warm a unit volume of soil 1°C. It requires about 582 cal/g to change liquid water to water vapor at 25°C; thus a greater proportion of heat received by the mulched soil is available for warming the soil. This tends to offset the effect of greater heat capacity and thermal conductivity. According to Van Wijk *et al.* (1959), the insulating effect and greater albedo of a mulch are much more important than the cooling effect of evaporation of soil water on bare soil as far as soil temperature is concerned.

CORN SEEDLING GROWTH AS RELATED TO TEMPERATURE

Duncan *et al.* (1972) in Florida measured air and growing point temperatures of corn during developmental stages of corn. While the growing point was below the soil surface, the mean of the maximum and minimum temperatures of the growing point was slightly higher than the mean of the maximum and minimum standard air temperatures of a nearby weather station when the soil moisture content was high. When the soil moisture content was low, the mean temperature of the growing point was 3.5°C higher than the mean air temperature. After the growing point emerged above the soil surface, the mean growing point temperature and mean standard weather station air temperature were approximately equal. The field air temperature (in canopy) during the day was higher than the standard air temperature in a nearby instrument shelter and growing point temperature. When the growing point was below the soil surface and for the days when the soil was classified as dry, the growing degree-days based on standard air temperatures averaged 25.7 degree-days/day while the growing degree-days based on growing point temperature averaged 34.4 degree-days/day. When the soil was classified as moist, the same calculations were 25.4 and 26.1 degree-days/day, respectively. Tassel emergence occurred 20 days after the growing point emerged above the soil surface. During this time, growing degree-days computed on the basis of growing point temperatures averaged only slightly below those computed in the normal way because the temperature of the growing point was about the same as measured in a standard instrument shelter.

According to Thompson (1966), the optimum mean daily air temperature (average of daily maximum and minimum air temperature) is 22–24°C (72–75°F) for June, July and August.

In Iowa, Van Wijk *et al.* (1959) computed growing degree-days based upon soil temperature at the 10-cm depth from May 7 to June 4 with and without 6.8 mt/ha of oat straw mulch spread uniformly over the soil surface; corn was planted on May 6. Accumulated growing degree-days computed on the basis of soil temperature at the 10-cm depth from May 7 to June 4 were 294 and 239 for non-mulched and mulched soils, respectively. Growth of corn as measured by dry matter accumulation was 19.0 and 10.3 g/10 plants for the non-mulched and mulched soils, respectively. Accumulated growing degree-days calculated on the same basis as above from May 7 to June 26 were 728 and 642 for the non-mulched and mulched soil, respectively. Dry matter production figures were 173 and 100 g/10 plants, respectively. The ratio of growth of the mulched to non-mulched increased only slightly when measured on June 19 as compared to June 4.

The data presented above by Duncan *et al.* (1972) and Van Wijk *et al.* (1959) indicate that the rate of corn growth and development during the first 30–45 days, or until the growing point emerges above the soil surface, is governed primarily by soil temperature surrounding the growing point. Since the presence of a mulch on the soil surface (whether it be no-tillage or not) lowers the average soil temperature in the surface few cm, the rate of growth of corn seedlings will be less on soil with a mulch (no-tillage) than the non-mulched soil (conventional tillage) if the soil temperature surrounding the growing point is less than the optimum temperature for corn growth. If, on the other hand, the soil temperature surrounding the growing point is higher than the optimum (which occurs frequently in the tropics and semi-tropics and less frequently even as far north as Lexington, Kentucky), then the rate of growth of corn seedlings would be greater on the mulched (no-tillage) than on the non-mulched soil (conventional tillage). Lehenbauer (1914) and Blacklow's (1972) growth rate curves of corn seedling shoots as a function of temperature are reproduced in Figure 2-1. Lehenbauer used an open pollinated genotype (Reed's yellow dent) and Blacklow used a hybrid genotype (UH 108). Both curves are essentially the same shape but the growth rate of the hybrid was approximately three times that of

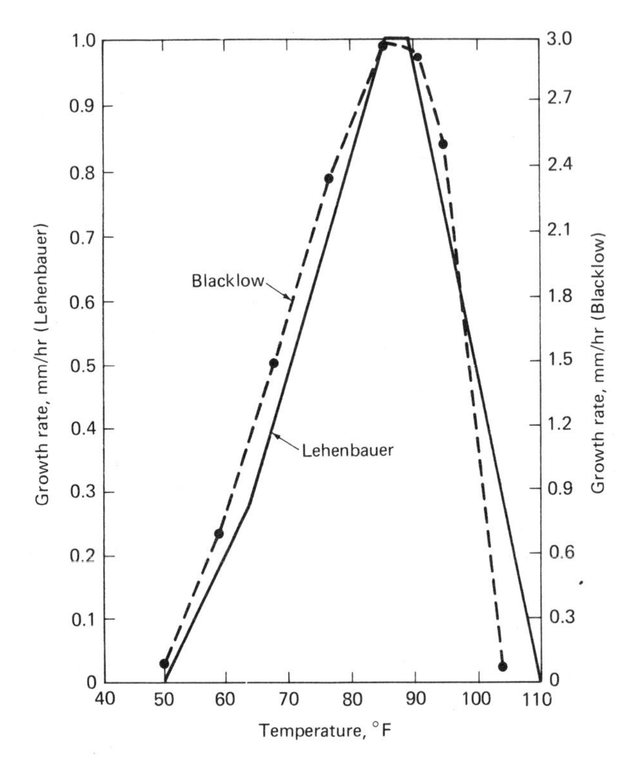

Figure 2-1. Lehenbauer's (1914) and Blacklow's (1972) growth rate curves of corn seedlings as a function of temperature.

the open pollinated genotype. Both grew the seedlings in the dark and measured growth rate soon after development of the shoot; however, Lehenbauer measured rate of corn seedling shoot growth after the shoots attained a length of 1.0–1.2 cm and Blacklow measured shoot growth rate beginning at about the stage of development as Lehenbauer but continuing over several days (five harvests). If the maximum temperature is less than optimum, the average of the maximum and minimum daily temperatures should estimate fairly accurately the growth rate of corn seedlings since the curve at temperatures below the optimum is essentially linear. However, if the minimum temperature is less than optimum and the maximum temperature is higher than optimum, then the average of the maximum and minimum daily temperature would be less accurate in predicting the actual growth of that day because the slope of the curve for greater than optimum temperatures is greater than the slope of the curve for less than optimum temperatures.

In the U.S. corn belt, cool soil temperatures due to a mulch (no-tillage) should not result in a reduction in grain yield as compared to non-mulched soil (conventional tillage) unless they delay maturity enough to cause frost damage and/or delay tasseling and silking enough to cause total photosynthesis during the grain filling period to be significantly reduced due to less radiation. Date of planting corn in Iowa, Illinois, Indiana and Ohio (the traditional U.S. corn belt) is at least as early as in Kentucky. Most corn in these states is planted during the month of May. For soils of similar color and volumetric soil water content, growing degree-days should correlate well with soil temperature surrounding the growing point of corn while it is below the soil surface and should predict with some degree of reliability regions where no-tillage can be expected to result in higher and lower corn yields.

The U.S. Weather Bureau's Weekly Weather and Crop Bulletin publishes accumulated weekly growing degree-days (GDD) computed from standard air temperature at 1.83 m above the soil surface from May 1 throughout the growing season. In regions where corn is grown, accumulated growing degree-days for the months of April, May and June can be used with some degree of reliability to predict if no-tillage can be practiced successfully at any location as far as the effect of temperature early in the growing season on growth of corn seedlings in concerned. Growing degree-days are calculated using the following equation.

$$\text{GDD} = [T_{\text{max}}\ (< 86°\text{F}) + T_{\text{min}}\ (> 50°\text{F})]/2 - 50°\text{F}$$

where

GDD = growing degree-days
T_{max} = maximum daily air temperature
T_{min} = minimum daily air temperature

Maximum temperatures above 86°F are entered as 86°F and minimum temperatures below 50°F are entered as 50°F. These restrictions are designed to take into account the differing responses of corn plants to low and high temperatures. At daily average temperatures below 50°F, rate of growth and development of corn is virtually zero.

Above 86°F, any increase in temperature does not result in an increase in rate of growth and development of corn. Temperatures above 86°F may be detrimental although it is implicitly implied by the equation that growth continues at the same rate as at 86°F. Growing degree-days calculated from soil temperatures would be more meaningful since the growing point of corn remains below the soil surface for about four to six weeks after planting, depending upon soil temperature and growing conditions. The mean standard air temperature is often less than the mean soil temperature surrounding the growing point during the first 30 days following planting for bare soil. Either high soil moisture and/or mulch would tend to lower the mean soil temperature during May in the corn belt. Low soil moisture would tend to increase soil temperature during May.

Growing degree-days, adjusted to 50°F base, accumulated from May 3 to May 17, May 17 to May 31, May 31 to June 14 and May 3 to June 14 for selected locations in U.S. (20-year average) are given in Table 2-3. These values were taken from Dye (1971), which were calculated from the mean of 20 years of standard air temperature data. Growing degree-days based upon growing point temperatures average 36 degree-days/day (50°F base) for optimum growth based upon Lehenbauer (1914) and Blacklow's (1972) corn growth curves. Duncan *et al.* (1972) reported an average of 34.4 and 26.1 degree-days/day calculated from growing point temperature (50°F base) for dry soil and moist soil, respectively, and that the growing point emerged above the soil surface 27 days after planting. The experiment was carried out in Florida, and during the course of it, the same plants were under dry soil conditions part of the time the moist soil conditions part of the time. The growing points emerged above the soil surface between 705 (27 days × 26.1 degree-days/day) and 929 (27 days × 34.4 degree-days/day) growing degree-days calculated from the average temperature of the growing point. The accumulated growing degree-days based upon the optimum temperature for corn (Lehenbauer) is probably less than 972 (27 days × 36 degree-days/day, assuming it requires less than 27 days for the growing point to emerge above the soil surface at the optimum growth rate). The average of Duncan's values is 817, which is the same order of magnitude as 972 and is probably biased upwards, since the maximum growing point temperature of the dry soil was about 99°F, well

Table 2-3. Mean growing degree-days, adjusted to 50° F base, accumulated from May 3 to May 17, May 17 to May 31, May 31 to June 14 and May 3 to June 14 for selected locations in U.S.; 20-year average. (From Dye, 1971.)

LOCATION	MAY 3 TO MAY 17	MAY 17 TO MAY 31	MAY 31 TO JUNE 14	TOTAL: MAY 3 TO JUNE 14
Des Moines, Iowa	169	201	268	638
Minneapolis, Minn.	133	163	233	529
North Platte, Nebr.	144	177	229	550
Omaha, Nebr.	186	220	285	691
Fargo, N.D.	116	154	208	478
Huron, S.D.	135	169	229	533
Green Bay, Wis.	109	140	199	448
Madison, Wis.	139	166	228	533
Fort Smith, Ark.	265	311	349	925
Dodge City, Kans.	192	231	289	712
Topeka, Kans.	205	244	301	750
Wichita, Kans.	217	261	319	797
Shreveport, La.	312	347	377	1036
Columbia, Mo.	205	242	304	751
St. Louis, Mo.	211	252	321	784
Oklahoma City, Okla.	244	288	338	870
Amarillo, Tex.	210	249	292	751
Corpus Christi, Tex.	371	398	414	1183
Fort Worth, Tex.	302	346	385	1033
Midland, Tex.	288	321	357	966
Chicago, Ill.	153	188	364	605
Moline, Ill.	168	200	269	637
Peoria, Ill.	170	206	278	654
Indianapolis, Ind.	175	209	278	662
Louisville, Ky.	211	255	307	773
Flint, Mich.	127	153	221	501
Grand Rapids, Mich.	136	161	235	532
Columbus, Ohio	170	207	267	644
Portland, Me.	87	123	166	376
Albany, N.Y.	134	165	224	523
Binghamton, N.Y.	108	134	193	435
Syracuse, N.Y.	126	157	220	503
Harrisburg, Pa.	168	204	269	641
Philadelphia, Pa.	173	206	274	653
Pittsburgh, Pa.	155	183	244	582
Williamsport, Pa.	154	179	238	571
Richmond, Va.	216	251	295	762
Macon, Ga.	291	335	358	984
Raleigh, N.C.	235	274	307	816
Memphis, Tenn.	276	322	361	959
Nashville, Tenn.	246	289	334	869

above the optimum of 86–90°F for corn growth. For temperatures at and below the optimum, growing degree-days calculated from growing point temperatures may well require the same accumulated growing degree-days for the growing point to emerge above the soil surface for a given genotype regardless of when it is planted if all growth factors other than growing point temperature remain constant, since the corn growth curve is essentially linear for these temperatures. Growing point temperatures throughout the U.S. corn belt would be below the optimum most of the time while the growing point is below the soil surface.

Growing degree days computed from soil temperatures at the 2.5-cm depth and temperature of the growing point while the growing point is below the soil surface will differ since the soil temperature at 2.5 cm will not be precisely the same as the temperature of the growing point. Soil temperature at the 2.5 cm soil depth is probably as representative of the actual temperature of the growing point as can be found in the literature since soil temperature data is rather scarce. Temperature data at the 5-cm and 10-cm soil depths are more prevalent than at the 2.5-cm depth. Soil temperatures at the 1–2.9-cm depth would probably be closer to the actual growing point temperature than at the 2.5-cm depth but such data is virtually nonexistent except for a few isolated cases. A comparison of growing degree-days computed from standard air temperature measurements and soil temperatures under a clipped, live, bluegrass sod and bare soil taken at the University of Kentucky and Ohio weather stations are given in Table 2-4. Growing degree-days computed from soil temperature under bluegrass sod are considerably less than for bare soil. Growing degree-days computed from standard air temperatures are in general less than those computed for bare soil temperatures for the Lexington location in April, May and June but the reverse is true for the Ohio locations in April. The values in Table 2-4 were calculated from monthly mean maximum and minimum temperatures using the equation of Dye (1971). The Lexington data are means of 13 years, 1967–1979, and there were 5 years from (1967–1979) that the average monthly maximum soil temperatures at the 2.5-cm depth for June were greater than 90°F. The growth rate of corn seedlings according to Lehenbauer and Blacklow's corn seedling growth curves decreases sharply above 90°F. A correction factor should be subtracted from

Table 2-4. Growing degree-days (50° F base) accumulated monthly from standard air temperature and from soil temperature under bare soil and under a clipped, living bluegrass sod for April, May and June.

LOCATION	MONTH	GROWING DEGREE DAYS COMPUTED FROM		
		STANDARD AIR TEMPERATURE	SOIL TEMPERATURE[1]	
			BARE	SOD
Lexington, Ky.	April	224	338	167
	May	396	564	486
	June	621	744	715
Coshocton, Ohio[2]	April	135	–	51
	May	329	–	326
	June	551	–	612
Ohio,[3] average of 14 locations	April	162[4]	63[5]	27[6]
	May	319	341	288
	June	531	639	567

[1]Soil temperatures were at soil depths of 2.5, 7.5 and 10 cm for Lexington, Coshocton and Ohio average, respectively.

[2]Personal communication, Edwards (1981).

[3]Personal communication, McIntyre (1981).

[4]Mean of total of 65 months for each month of April, May and June.

[5]Mean of total of 46, 51 and 53 months for April, May and June, respectively.

[6]Mean of total of 44, 45 and 46 months for April, May and June, respectively.

growing degree days when the temperature exceeds 90°F when considering Lehenbauer and Blacklow's corn seedling growth curves. When using the same principle above 90°F as is used to calculate growing degree days between 50°F and 86°F, one gets a correction factor of

$$\text{Corrected GDD} = -\ [(90°\text{F} + T_{max})/2 - 90°\text{F}]$$
$$= -\ [T_{max} - 90°\text{F}]/2.$$

If T_{max} exceeds 110°F, it would be entered as 110°F. If T_{max} is $\leq$ 90°F, the correction factor would be zero. When subtracting this corrected GDD for the bare soil in May and June, one gets corrected GDDs of 545 and 695 for the months of May and June, respectively, for the Lexington location. Then the corrected GDDs in Table 2-4 for the bare soil for the months of May and June are 545 and 695, respectively, instead of 564 and 744, respectively. The correction factor for the GDDs presented in Table 2-4 for standard air temperature and for sod is zero, since those maximum temperatures did not exceed 90°F for those years at the Lexington location (1967–1979). The average percent increase of GDDs reported in Table 2-4 for the months April, May and June, of the bare and sod soil temperatures, compared to those computed from standard air temperatures, is 33 and 10 percent, respectively.

Near Lexington, Kentucky, most of the corn is planted about May 15. Assuming that the growing point of corn will emerge above the soil surface 30 days after planting at Lexington, the GDDs as computed from soil temperatures at the 2.5-cm depth are approximately 600 and 650 for no-tillage and conventional tillage, respectively, during the 30 days following planting, if planted on May 15. This calculation assumes the soil temperature regime for no-tillage and conventional tillage to be the same as under bare soil and under sod.

The above discussion comparing GDDs when computed from standard air temperatures and soil temperatures was given because standard air temperatures are readily available nearby for every location in the U.S. but soil temperatures are not.

Assuming that the soil temperatures at the 2.5 cm depth under bare soil is representative of the growing point of corn seedlings on conventional tillage and the soil temperature at the 2.5 cm depth under a living, clipped bluegrass sod is representative of the growing point temperature of corn seedlings on no-tillage, one can graph the approximate growth rate of a corn seedling as governed by temperature during a 24-hr period using Lehenbauer and Blacklow's growth rate curves for corn seedlings. Such graphs are shown in Figures 2-2, 2-3 and 2-4 for an average day in April, May and June at Lexington, Kentucky. These curves were constructed using the mean monthly maximum and minimum temperatures at the 2.5 cm soil depth (mean of 13 years). Growth rates at the mean monthly maximum and minimum temperatures were taken from Lehenbauer and Blacklow's corn seedling growth curve to establish two points. The remainder of the curves were based upon the daily soil temperature curve being sinusoidal. The growth rate at midnight was assumed to be mean of growth rate at the maximum and minimum soil temperatures. The maximum and minimum soil temperatures were assumed

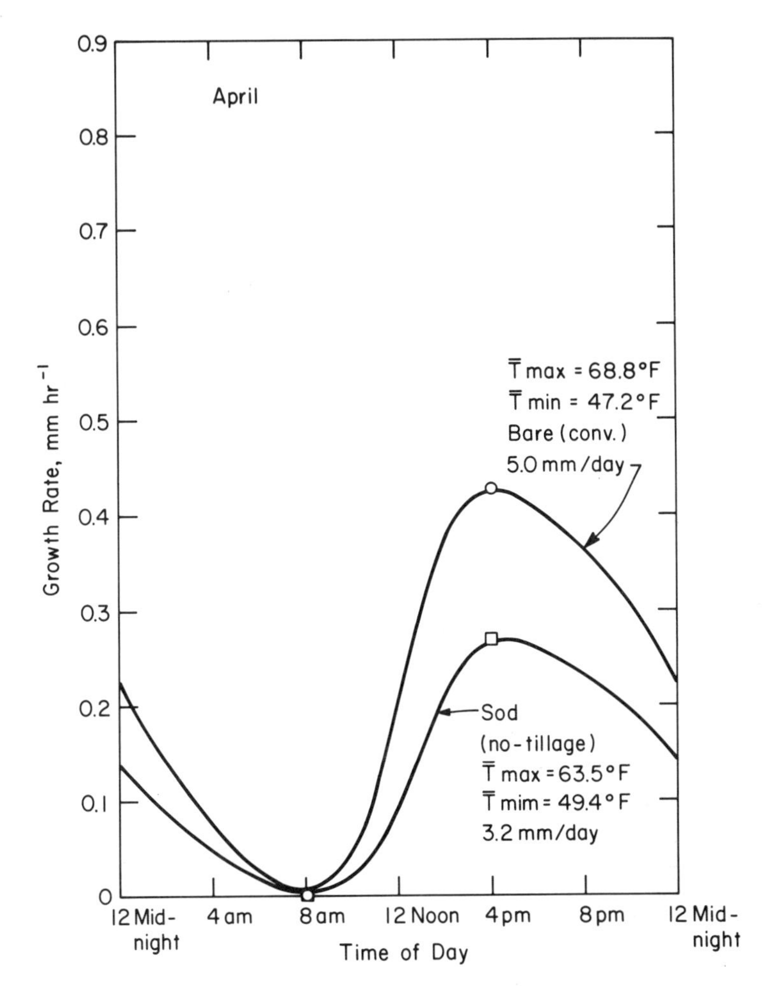

Figure 2-2. Growth rate of corn seedlings during a typical day in April at Lexington, Kentucky for bare (conventional tillage) and sod (no-tillage).

to occur at 4 p.m. and 8 a.m., respectively. Therefore, the growth rate curves are not exact but a general idea of the change in the growth rate throughout the day can be obtained from these figures. Note that the growth rates of the graphs are those from Lehenbauer's curve; to convert to Blacklow's growth rates multiply by 3.

As can be seen from Figure 2-2, very little growth of corn seedlings would be expected in April. Integrating the area under each curve and taking the ratio of total growth for the day of no-tillage (sod)

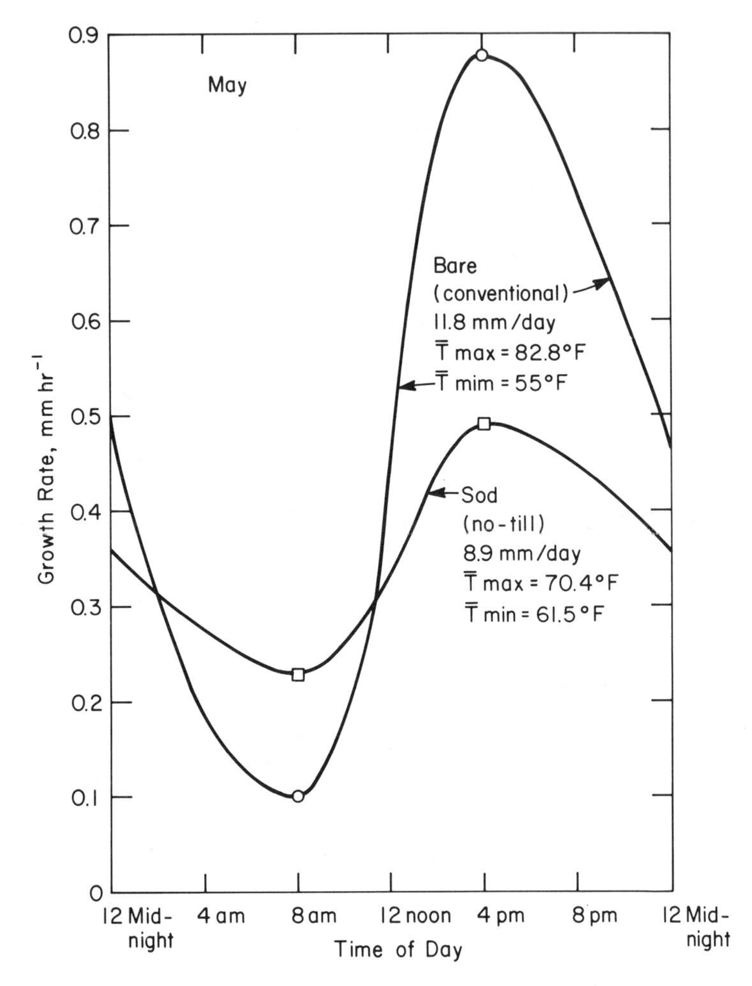

Figure 2-3. Growth rate of corn seedlings during a typical day in May at Lexington, Kentucky for bare (conventional tillage) and sod (no-tillage).

and conventional tillage (bare) gives a value of 0.64. The growth rates in May are almost two times as great as in April (see Figure 2-3). The ratio of growth for May of no-tillage to conventional tillage is 0.75. Figure 2-4 shows a temporary dip in the curve after reaching a maximum growth rate of 1 mm hr^{-1} at about 2:00 p.m. for the conventional tillage (bare) curve. This is caused by the maximum soil temperature going above the optimum temperature

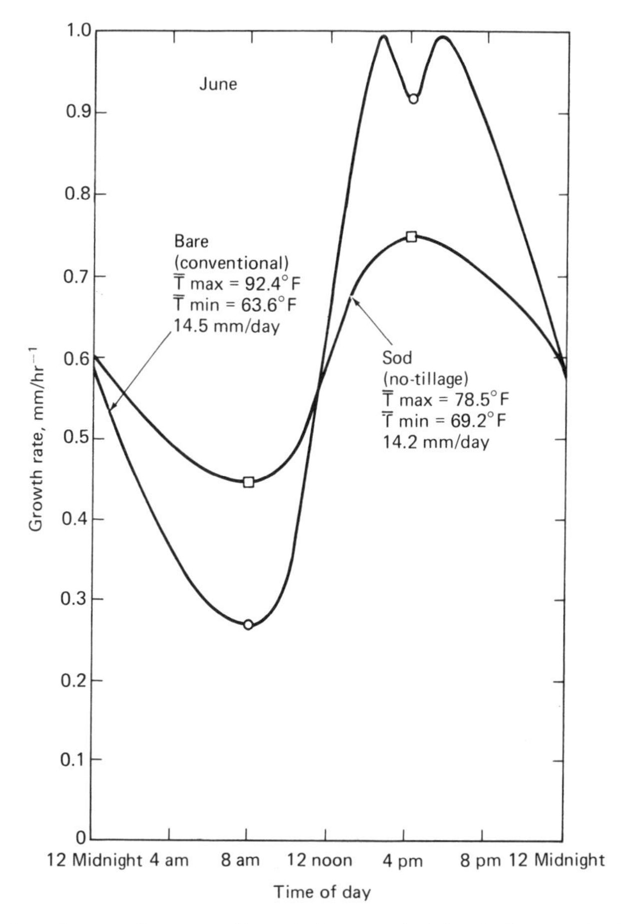

Figure 2-4. Growth rate of corn seedlings during a typical day in June at Lexington, Kentucky for bare (conventional tillage) and sod (no-tillage).

(86–90°F) for growth of corn seedlings. Integrating under the two curves as before, the ratio of total growth for the day of no-tillage to conventional tillage is 0.98.

During the month of May, 1977 at Lexington, Kentucky, an especially warm May, the mean monthly maximum and minimum soil temperatures at the 2.5-cm depth for conventional tillage (bare) soil were 104°F and 61°F, respectively. The same respective soil temperatures were 78°F and 70°F for no-tillage (sod). Figure 2-5 shows the growth rate of corn seedlings during an average day in May

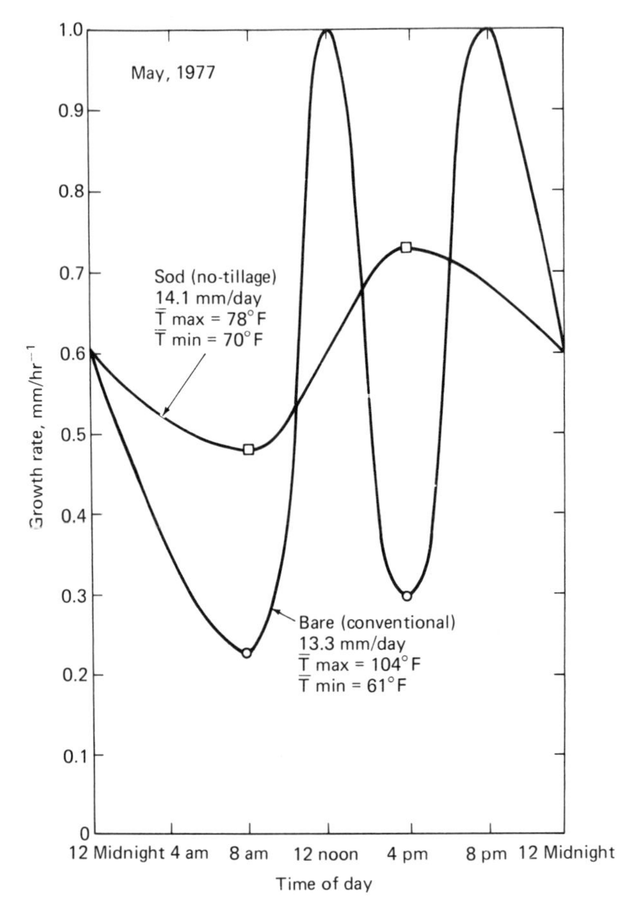

Figure 2-5. Growth rate of corn seedlings during a typical day in May, 1977 at Lexington, Kentucky for bare (conventional tillage) and sod (no-tillage).

1977. There is an extreme fluctuation of growth rate for conventional tillage throughout the day. Growth rates reached minima and maxima two times during the 24-hr period for the conventional tillage (bare). The first minimum occurred at about 8 a.m. when the minimum soil temperature occurred; the second minimum occurred at about 4 p.m., when the maximum soil temperature of 104°F occurred. The first maximum occurred at about 12 noon, when the soil temperature reached 86°F; the second maximum occurred at about 8 p.m., when the soil temperature decreased to

90°F. There is much less fluctuation in the curve for no-tillage (sod). I do not know whether there is a deleterious effect on corn seedling growth physiologically due to extreme fluctuations as shown in Figure 2-5 for bare soil. The ratio of total growth of the no-tillage (sod) to conventional tillage (bare) is 1.06 in Figure 2-5. During May of 1977, the total growth of corn seedlings on no-tillage (sod) was favored slightly over that of conventional tillage. The mean monthly maximum standard air temperature in May 1977 was 80°F while the maximum soil temperature for bare soil at the 2.5-cm depth was 104°F. The reasons for these high soil temperatures as compared to May, in general, at Lexington were that little or no rainfall fell from mid-April to May 30, causing the soil to be much drier at the 2.5 cm depth than normal, and the mean maximum standard air temperature was 7°F above normal.

In most years, most of the corn is planted during the month of May in the corn belt (Iowa, Illinois, Indiana, Ohio, Michigan and Nebraska) of the U.S. Since soil temperatures under mulched soil, including no-tilled soil with a mulch, are lower than for conventional tillage, germination and early growth of the corn seedlings are slower under mulched soil than conventionally tilled soil. The slower growth of corn seedlings due to the cooler soil temperatures will not significantly reduce grain yields unless silking and tasseling are significantly delayed due to less total photosynthesis (less total net radiation) during the grain filling period and/or frost damage before maturity. In general, in most years on well-drained, medium textured soils, corn grain yields on no-tillage would be expected to be equal to or greater than corn grain yields on conventional tillage (both planted at the same time) south of a line from New York City to Philadelphia, to Harrisburg, Pennsylvania across northern West Virginia, across the southern half of Ohio, Indiana and Illinois, across the southern fourth of Iowa and the southeastern portion of Nebraska. In warmer and/or drier years and in cooler and/or wetter years, this line would move north or south. Just how far north or south it would be extended would depend upon the magnitude of increase or decrease in soil temperatures. Areas where soil properties cause a decrease or increase in soil temperature would be excluded from the above generalization. For example, poorly-drained soils in the corn belt in May have a surface soil temperature lower than do well-drained soils due to increased thermal conductivity and higher heat capacity. Likewise, sandy, well-drained soils have a surface soil temperature higher than do medium-textured, well-drained soils. Sandy soils generally have less water than do well-drained soils and, therefore, the heat capacity is lower, On the other hand, sandy soils generally have a higher albedo than medium-textured soils but it is not enough higher to outweigh the lower thermal conductivity and lower heat capacity. Aspect on the same soil can have a significant influence on soil temperatures as well.

In Table 2-3, the isoline of mean growing degree days (based upon standard air temperatures) accumulated from May 3 to May 31 of the boundary of the line mentioned above for success of no-tillage as far as corn grain yields are concerned is approximately 375 growing degree-days. Using the percent increase in growing degree-days computed from soil temperatures under a bare surface and under sod over that computed from standard air temperatures for May of 45 and 23 percent, respectively, at Lexington, would mean about 545 and 460 growing degree-days during May for bare (conventional tillage) and sod (no-tillage), respectively; the Lexington data only from Table 2-4 were used in this calculation since soil temperatures were taken at the 2.5-cm depth. These soil temperatures at Lexington, Kentucky were measured in a medium-textured (silt loam) well-drained soil.

TIME OF PLANTING

Time of planting has an effect upon ultimate corn-grain yields in the U.S. cornbelt. Yields decrease on the average approximately 6–7 percent for each week that planting is delayed after May 10. This is shown in Figure 2-6, where percent of maximum yield is plotted as a function of planting date. The data in Figure 2-6 are from Michigan (Aldrich *et al.*, 1975), Illinois, (*Illinois Agronomy Handbook*, 1978), Iowa (*A Successful Farming Book, Soil and Crops*, 1967) and Kentucky. The scatter of the data is surprisingly small when considering that data from four states are plotted in the figure. Linear regression was run for percent of maximum yield against planting date. The equation for this regression was $Y = 107.2 - 0.97X$, where Y = percent of maximum yield and X = days after May 10.

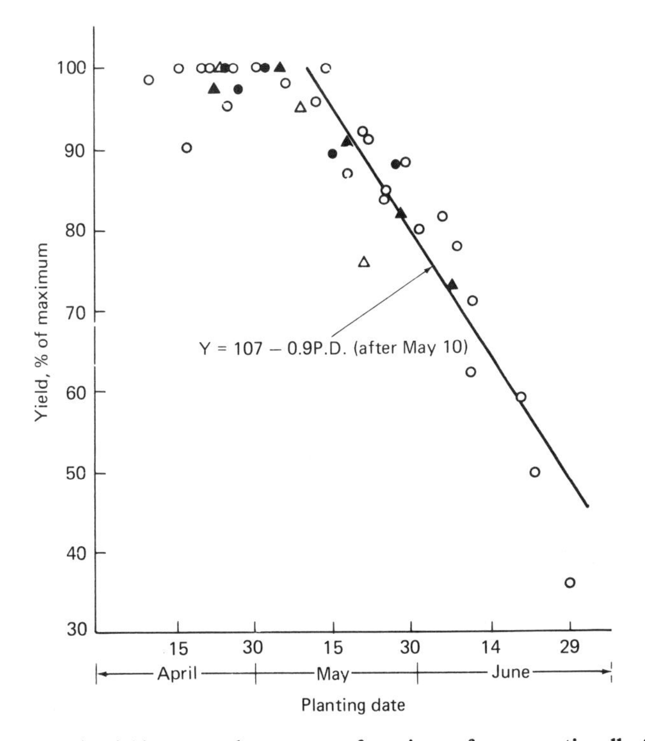

Figure 2-6. Corn grain yield expressed as percent of maximum for conventionally tilled corn as a function of planting date: o–Lexington, Kentucky, 1-3-year average; *x*–Michigan, 10-year average (adapted from Aldrich *et al.*, *Modern Corn Production*, 1975); Δ–Iowa, 1-year average (reprinted from *Successful Farming Soils and Crops Book*, copyright Meredith Corporation 1967, all rights reserved); •–Illinois, 3-year average (from *Illinois Agronomy Handbook*, 1978).

Figure 2-7 shows the percent of corn planted in Kentucky, Iowa and the 17 major corn producing states during the month of May in the U.S. (Murphy, 1981). Approximately 47, 54 and 60 percent of corn is planted by May 10 in the 17 major corn producing states, Kentucky and Iowa, respectively. This means, on the average, that approximately 50 percent of the corn is planted when maximum yields cannot be obtained.

Figure 2-8 shows percent of maximum yield of no-tillage corn grown in central Kentucky as a function of planting date (Bitzer, 1980 and Phillips, 1980). These data cannot be used with much confidence since each data point is for one year only. It would

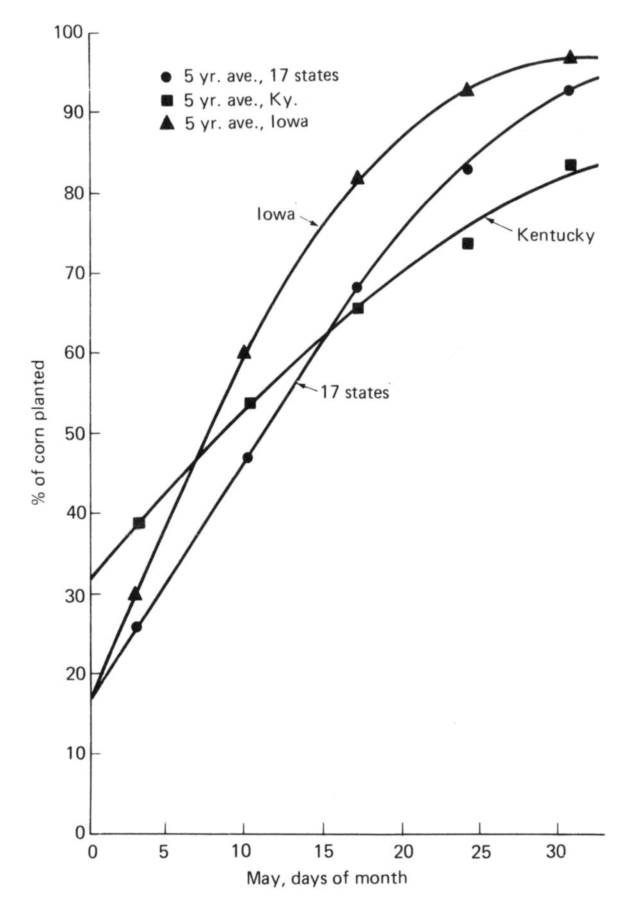

Figure 2-7. Percent of corn planted during the month of May (from R. M. Murphy, 1981).

appear that the optimum planting date for no-tillage corn is no later than for conventional tillage (see Figure 2-7). Another interesting indication is that corn planted in April in central Kentucky under no-tillage results in a reduction of yield; corn planted conventionally does not show this. This slope of the linear regression indicates that corn grain yield is reduced by about 0.44 percent each day planting is delayed after May 10–15; the comparable reduction for conventional tillage, Figure 2-7, is 0.92 percent for each day planting is delayed after May 10. Again it should be pointed out that the data presented in Figure 2-8 for no-tillage is based upon

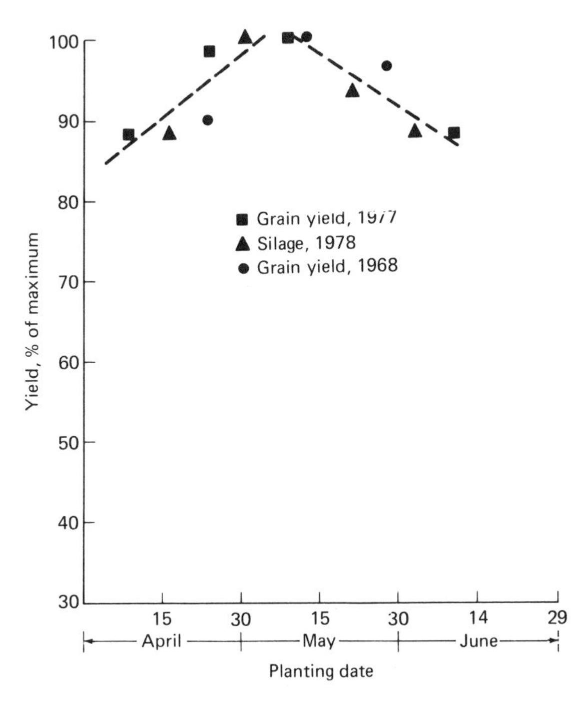

Figure 2-8. Corn grain yield expressed as a percent of maximum for no-tillage as a function of planting data: o–grain yield, 1977; *x*–sillage, 1978; o–grain yield, 1968 (from Bitzer, 1980 and Phillips, 1980).

limited information; however, much confidence can be placed upon the data presented in Figure 2-7 for conventional tillage. In Figure 2-8 the positive slope of the data points during April, if a positive slope does in fact exist, could be due to the effect of low soil temperatures during early growth of the seedling on final grain yield. As pointed out above, the reduction in grain yield appears to be greater for conventional tillage than for no-tillage for corn planted after May 10–15. Some of this difference in conventional tillage and no-tillage can be attributed to less water stress in no-tillage than in conventional tillage. However, a majority of this difference cannot be attributed to water-stress. Much of it can be attributed to lower net radiation and other environmental factors during the grain-filling period. Corn planted on May 1, May 15, June 1 and June 15 will silk on or about July 15, July 23, August 6 and August 17, respectively. The respective maturity dates will be on or about September 10, September

Table 2-5. Mean daily solar radiation (1950–1962), ly/day, by month for selected locations in the U.S. (From Bennett, 1965.)

MONTH	LOCATION					
	BISMARCK, N.D.	DES MOINES, IOWA	COLUMBIA, MO.	COLUMBUS, OHIO	LOUISVILLE, KY.	ATLANTA, GA.
January	157	168	183	154	164	238
February	242	237	256	217	227	307
March	341	322	344	314	325	390
April	450	425	434	406	420	522
May	544	514	532	517	515	586
June	598	568	578	566	560	586
July	608	568	577	563	550	562
August	516	500	528	506	498	531
September	386	405	439	412	408	441
October	272	301	323	296	303	379
November	162	189	217	174	190	290
December	126	138	168	134	150	213
Average for year	367	361	382	355	359	420
Average for April–September	517	497	515	495	492	538

25, October 10 and October 25. Mean daily solar radiation drops off sharply in September and October as compared to June, July and August (see Table 2-5).

The mean daily solar radiation for September is approximately 100 ly/day less than in August for all locations shown in Table 2-5. Likewise, the values are approximately 100 ly/day less for October than in September. This difference translates to approximately 20 percent less photosynthesis for September as compared to August and approximately 40 percent less for October as compared to August. There is not much difference in mean daily solar radiation among the locations shown in Table 2-5 for the growing season, April through September, as well as for the complete year if Atlanta, Georgia is excluded. There is, however, much difference in temperature during the fall, winter and early spring months, but the mean

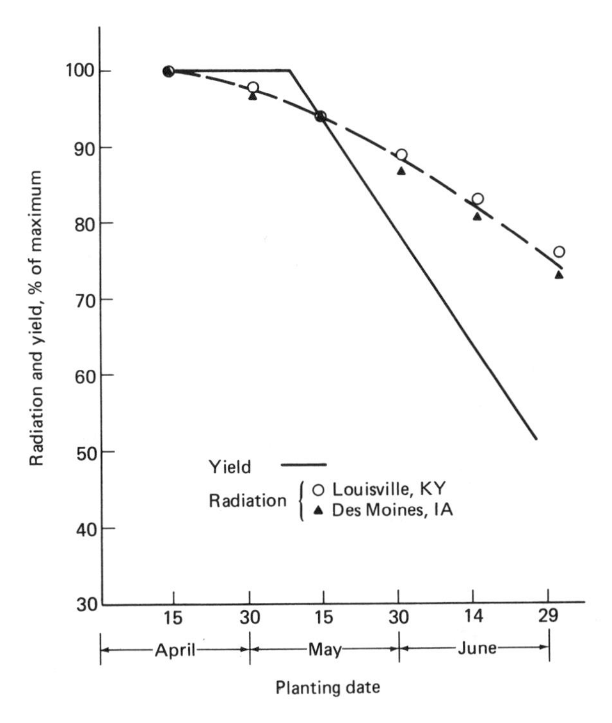

Figure 2-9. Relationship of corn grain yield and date of planting and solar radiation for a period of 15 days following silking corresponding to planting dates shown, planting date of April 15 = 100% at Lexington, Kentucky and Des Moines, Iowa.

daily temperature is not much different from May 1 through September 30. The average daily air temperature for Des Moines, Iowa is 20.5°C and 21.9°C for Lexington, Kentucky from May 1 through September 30 (Mitchell, 1978).

The best fit of data presented in Figure 2-6 is reproduced in Figure 2-9, where percent of maximum yield as a function of planting date is plotted. Superimposed upon that is percent of maximum of net solar radiation at Louisville, Kentucky and Des Moines, Iowa for 15 days following silking corresponding to planting dates of April 15, April 30, May 15, May 30, June 15 and June 30. As can be seen in Figure 2-9, solar radiation during the grain filling period correlates well with grain yield for the planting dates. Other factors such as hot, drying winds during pollen shed, soil water, early frosts, insects, and diseases appear to have about an equal effect upon grain yield in the U.S. cornbelt as to net solar radiation during the grain filling period.

REFERENCES

Aldrich, S. R., W. R. Scott, and E. R. Leng. 1975. *Modern Corn Production,* 2nd Edition, Chapter 5. A & L Publications, Inc., F. and W. Publishing Corp., Cincinnati, Ohio.

Bennett, Iven. 1965. Monthly maps of mean daily insolation for the United States. *J. Solar Energy* **9**:145–158.

Bennett, O. L., E. L. Mathais, and P. E. Lundberg. 1973. Crop responses to no-till management practices on hilly terrain. *Agron. J.* **65**:488–491.

Bitzer, M. J. 1980. Personal Communication. Extension Professor, Department of Agronomy, University of Kentucky, Lexington.

Blacklow, W. M. 1972. Influence of temperature and elongation of the radicle and shoot of corn (*Zea mays* L.). *Crop Sci.* **12**:647–650.

Burrows, W. C. and W. E. Larson. 1962. Effect of amount of mulch on soil temperature and early growth of corn. *Agron. J.* **54**:19–23.

Denny, Lyle. 1978. *National Oceanic and Atmospheric Administration. Climatological Data: National Summary* **29**. (Mitchell, D. G., Director of the National Climatic Center.)

Duncan, W. G., D. R. David, and W. H. Chapman. 1972. Developmental temperatures in corn. *Soil and Crop Sci. Soc. of Florida, Proc.* **32**:59–63.

Dye, Lucius W., Editor. 1971. *U.S. Weather Bureau: Weekly Weather and Crop Bulletin* **58** (March 29 issue).

Editors of Successful Farming. 1967. *Successful Farming Soils and Crops Book,* 11th Edition, p. 26. Meredith Publishing Co., Des Moines, Iowa.

Edwards, W. M. 1981. Personal Communication. USDA–SEA, Coshocton, Ohio.
Illinois Agronomy Handbook. 1978. University of Illinois, Circ. No. 1129.
Lal, Rattan. 1974a. Soil temperature, soil moisture and maize yield from mulched and unmulched tropical soils. *Pl. and Soil* **40**:129–143.
Lal, Rattan. 1974b. No-tillage effects on soil properties and maize production in Western Nigeria. *Pl. and Soil* **40**:321–331.
Lehenbauer, P. A. 1914. Growth of maize seedlings in relation to temperature. *Physiol. Res.* **1**:247–288.
Mitchell, D. B. 1978. *Climatological Data: National Summary, National Oceanic and Atmospheric Administration* **29.**
Moody, J. E., J. N. Jones, Jr., and J. H. Lillard. 1963. Influence of straw mulch on soil moisture, soil temperature and the growth of corn. *Soil Sci. Soc. Am. Proc.* **27**:700–703.
McIntyre, J. M. 1981. Personal Communication. Agricultural Meteorologist, Mideast Agricultural Weather Service Center, Purdue University, West Lafayette, Indiana.
Murphy, R. M. 1981. Unpublished Data. Assistant Statistician in Charge, Kentucky Crop and Livestock Reporting Service.
Papadakis, J. 1961. *Climatic Tables For the World,* Copyright 1961 by J. Papadakis, Buenos Aires, pp. 55–142.
Phillips, S. H. 1980. Personal Communication. Associate Director of Agricultural Extension, College of Agriculture, University of Kentucky, Lexington.
Thompson, Louis M. 1966. *Weather Variability and the Need for a Food Reserve.* The Center for Agricultural and Economic Development, College of Agriculture, Iowa State University of Science and Technology. CAED Report 26.
Todd, David Keith, Editor. 1970. *The Water Encyclopedia.* Water Information Center, Inc., Syosset, N.Y.
Van Wijk, W. R., Larson, W. E., and Burrows, W. C. 1959. Soil temperature and the early growth of corn from mulched and unmulched soil. *Soil Sci. Soc. Am. Proc.* **23**:428–434.
Walker, John M. 1970. Effects of alternating versus constant soil temperatures on maize seedling growth. *Soil Sci. Soc. Am. Proc.* **34**:889–892.
Willis, W. O. and Amemiya, M. 1973. Tillage management principles. Soil temperature effects. *Proc. National Conserv. Tillage Conference,* Des Moines, Iowa, pp. 22–42.
Willis, W. O., Larson, W. E., and Kirkham, D. 1957. Corn growth as affected by soil temperature and mulch. *Agron. J.* **49**:323–328.

3
Soil Adaptability for No-Tillage

Robert L. Blevins
Professor of Agronomy
University of Kentucky

SOME FACTORS AFFECTING ADOPTION OF NO-TILLAGE

The unique properties of individual soils determine their limitations and suitability for land use. Because of these wide ranges of differences among soils of the world it is understandable that no one tillage system would be best suited for all soils. Soil properties and climatic conditions should be carefully evaluated before selecting a tillage system. Soil conditions that favor no-tillage farming or modifications of these systems will be discussed in this chapter as well as conditions that are less favorable for the adoption of no-tillage farming techniques.

Acceptance and subsequent adoption of no-tillage corn production systems has been most successful on well-drained to moderately well-drained, medium-textured soils. Soils that are potentially droughty for short periods of the growing season usually respond very favorably to no-tillage management when compared to conventional tillage. When making soil management decisions on specific soils one should determine whether there is a need for erosion control, potential moisture storage and supplying capacity of the soil, and the internal drainage characteristics of the soil. For example, on nearly level, poorly-drained soil, erosion is no longer a hazard to be concerned with and excessive water may be the major limitation to deal with. In the case of nearly level, wet soils, leaving residues at the soil surface and no-tilling may in effect increase the wetness hazard. The surface mulch reduces soil water evaporation, which makes the soil warm up more slowly in the spring. For this reason farmers on the nearly level, imperfectly drained soils of the cornbelt region of the U.S. have shown less interest in using no-tillage as compared to more ideal landscapes for no-tillage that have sloping land and good internal drainage.

No-tillage farming is a relatively new concept made possible through the development of chemical herbicides that can provide good weed control without using tillage. The interest in and adoption of no-tillage systems of crop production is rapidly growing in the U.S. (Figure 3-1) and throughout the world. Data published by *No-Till Farmer* magazine (Lessiter, 1982) presents an overview of no-tillage in the U.S. by crop production regions during 1982. The number of hectares in no-tillage production in 1972 was 1,342,000; in 1982 this increased to 4,198,000 hectares, or about a four-fold increase in a decade. It is especially encouraging that areas that are rapidly adopting no-tillage occur in the eastern half of the U.S., where soil erosion by water is the major soil hazard and limitation. The no-tillage system of row crop production is especially well-adapted

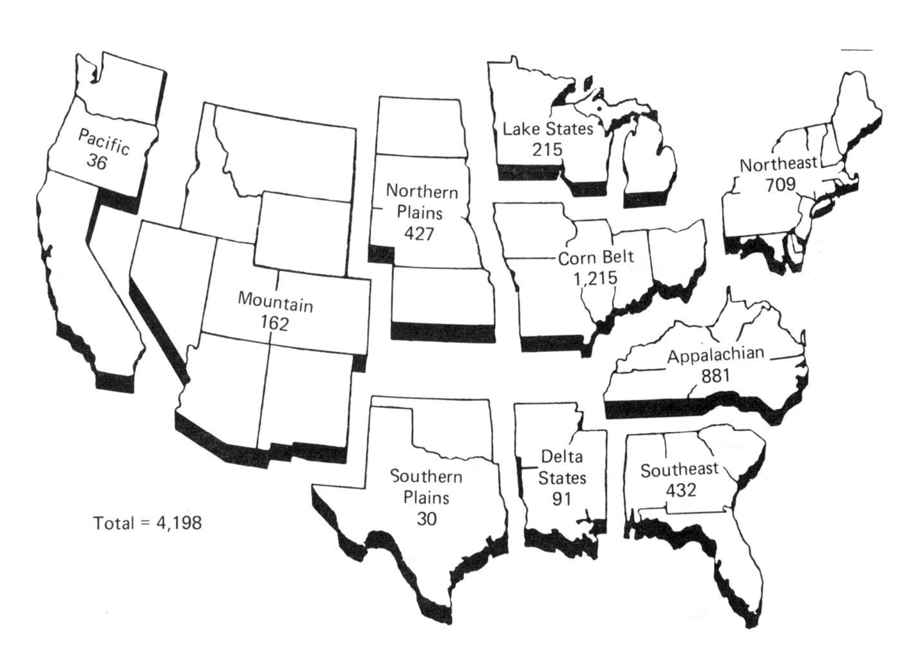

Figure 3-1. Land in no-tillage systems by crop production region (thousands of hectares) of the U.S. in 1982 (from Lessiter, 1982).

to sloping soils with silt loam surface textures such as those in the Appalachian region. The southern sections of Illinois, Indiana and Ohio are also characterized by sloping land, and no-tillage is particularly well-suited to these sectors, as well as the hilly loessal soils in Iowa and Missouri.

Because of the soil and water conservation benefits of a no-tillage system, it would seem that this system is suitable over a wide range of soil and slope conditions normally used for grain production.

A comparison of corn grain yield of no-tillage and conventional tillage is given in Table 3-1. This comparison includes only well- to moderately well-drained soils. The higher yields for no-tillage appear to be closely related to more efficient use of soil water. Only the trials on the Faywood soil and moderately well-drained Grenada silt loam soil with a fragipan at the 60-cm depth showed no yield advantage for using no-tillage over conventional tillage.

Table 3-1. Comparison of corn yields on different soils using no-tillage versus conventional tillage methods. (From Blevins *et al.*, 1971.)

SOIL	SOIL TEXTURE	SOIL SLOPE (%)	PARENT MATERIAL	CORN YIELD (kg/ha)	
				NO-TILL	CONVENTIONAL
			1969		
Crider	Silt loam	3	Loess over limestone	8,969	8,028
Donerail	Silt loam	3	Phosphatic limestone	8,530	7,338
Faywood	Silty clay loam	7	Limestone	8,279	8,342
Grenada	Silt loam	2	Loess over acid sandstone and shale	6,523	6,523
Loradale	Silt loam	6	Phosphatic limestone	8,154	6,899
Lowell	Silt loam	8	Limestone	9,345	8,342
			Average	8,279	7,589
			1970		
Loradale	Silt loam	6	Phosphatic limestone	7,275	7,213
Maury	Silt loam	3	Phosphatic limestone	6,523	5,645
Shelbyville	Silt loam	4	Limestone and calcareous shale	8,593	7,965
Tilsit	Silt loam	2	Acid sandstone and shale	6,460	3,933
			Average	7,213	6,460

Although these reports present a very positive story favoring no-tillage, questions remain concerning the adaptability of poorly-drained soils to no-tillage management and how climatic conditions influence your choice in selecting the best tillage system (see Chapter 2).

SOIL DRAINAGE

Soils with medium texture and good internal drainage characteristics respond very favorably to no-tillage management. Numerous researchers (Blevins *et al.*, 1971; Moschler *et al.*, 1973; Bone *et al.*, 1976) all have reported higher yields using no-tillage. No-tillage production of grain crops has been used successfully on well- to excessively well-drained sandy soils such as the loamy sands in Delaware (Mitchell and Teel, 1977) and on sandy loam and sandy clay loam soils in Georgia (Gallaher, 1977). The no-tillage system allows more efficient use of stored soil water during short-term droughts which occur frequently in the coarse-textured Coastal Plain soils of the eastern seaboard of the U.S.

In contrast, soils showing the poorest response to no-tillage management are usually those with slow internal drainage properties and nearly level slopes. The relationship of soil drainage and texture is well-illustrated in Table 3-2. The medium-textured, silt loam soils produced better corn yields for no-tillage as compared to the finer-

Table 3-2. Corn yield by tillage systems on major Ohio soils. (From Bone *et al.*, 1976.)

SOIL	SOIL TEXTURE	DRAINAGE CLASS	YIELD, BY TILLAGE SYSTEM (kg/ha)		
			CONVENTIONAL	MINIMUM	NO-TILLAGE
Wooster	sil	Well-drained	6,967	7,156	8,160
Rossmoyne	sil	Moderately well-drained	8,913	8,725	9,039
Crosby	sil	Somewhat poorly-drained	8,599	9,039	8,976
Brookston	sicl	Very poorly-drained	10,168	9,541	9,164
Hoytville	sicl	Very poorly-drained	8,474	8,411	7,156

textured, wet soils. The well-drained Wooster soil showed the greatest yield increase over conventionally tilled comparison plots. This difference between tillage treatment decreased as degree of soil wetness increased. The more clayey, very poorly-drained Brookston and Hoytville soils produced lower yields when managed as a no-tillage system. These results agree with similar findings reported in Indiana (Griffith *et al.*, 1973), on reduced yields from no-tillage on wet clayey soils. Griffith *et al.* (1973) concluded that no-tillage (coulter) corn production was not well-suited as a management system on wet, clayey soils of central and northern Indiana. However, in southern Indiana on more sloping land and better-drained soils (Bedford silt loam) the no-plow systems can produce yields equal to or higher than conventional tillage.

The soils of Britain have been classified by Cannell *et al.* (1978), as to suitability for repeated direct drilling (no-tillage) of cereal crops. The suitability map shows that limestone and chalk and other well-drained loam soils are equally suited for spring and winter crops. Field experience indicates that soils with slow internal drainage and periodic waterlogging conditions are not well-suited to .direct drilling. The soils in Britain with the most favorable properties are usually well- and moderately well-drained soils with a generally loamy soil texture and may include some humose clayey soils. The least favorable group of soils for direct drilling includes excessively drained soils and soils affected by fluctuating ground water and/or flooding that are often associated with sandy and silty soils with a low organic matter content. Compaction is also a problem for these soils and tillage may be required to ameliorate these conditions.

Reasons for reduced no-tillage yields on wetter soils are not fully understood but several factors have been identified that cause a negative response in plant growth. Indiana researchers concluded that germination and weed control is more of a problem in no-plow tillage systems as compared to conventional and these problems often are more severe on wet fine-textured soils. On wet, clayey soils in northern Ohio (Van Doren *et al.*, 1976), more corn root damage was observed in no-tillage plots as compared to plowed treatments, especially if the area had been in continuous corn. This root damage was associated with the higher incidence and activity of *Pythium gramicola.* Perhaps more organisms such as *pythium* are present when associated with continuous corn production on soils that have frequent periods of near water saturation early in the growing season.

No-tillage corn production on imperfectly drained soils in Kentucky has been less successful than conventional tillage and more refined management inputs are required (Murdock, 1974). Suggested reasons for the poorer corn yield response and lower plant populations on wet soils are: (1) poor growth and emergence related to lower soil temperatures; (2) loss of nitrogen; (3) disease damage; (4) insect damage to plants during a time of reduced vigor; and (5) poor growth due to low levels of oxygen following rainy periods.

The wetter the soil the slower it warms in the spring. This principle may pose a problem in more northern latitudes. The no-tillage system where mulches are left at the surface, reduce evaporation from the soil surface and acts as an insulation or barrier to the absorption of solar radiation. The cooler environment under no-tillage as compared to plowed soil is less favorable for rapid and uniform germination in cooler climatic regions. Cool soil and air conditions result in slower and less vigorous growth of the seedling. This reduced plant vigor results in a situation where the plants are more susceptible to disease or insect damage. The wetter soil condition may reduce the amount of total oxygen available at any one time and result in denitrification and subsequent loss of nitrogen from the soil.

Studies by Murdock and Blevins (1981) in Kentucky on no-tillage corn grown on Tilsit silt loam, a moderately well-drained soil, underlain by a fragipan at 45–55-cm depth showed that early planting dates produce lower yields and reduced stands for no-tillage as compared to conventional tillage (Table 3-3). Higher yields were observed for no-tillage than for conventional tillage at the later date when cooler soil temperatures and periodic excessive wetness were no longer a problem. Preliminary studies at Kentucky show a good response from the use of fungicides at corn planting time for no-tillage management of wet soils.

Poorly-drained soils which dry out more slowly in the spring usually occur on nearly level slopes where surface runoff and erosion are of little concern. Thus, the advantage that no-tillage offers in erosion control and surface runoff is not a major concern in making choices in tillage systems to use. If soils with slow internal drainage are managed as a no-tillage system, the management must be different than conventional tillage. A higher level of management will be required if no-tillage is used on these wetter soils. Planting dates will

Table 3-3. Effect of planting dates on corn yields and plant population for no-tillage and conventionally tilled treatments on a moderate to somewhat poorly drained soil in western Kentucky (three-year aveage). (From Murdock and Blevins, 1981.)

PLANTING DATE	TILLAGE SYSTEM	YIELD (kg/ha)	PLANT POPULATION/ha
1st (May 1)[1]	No-tillage	7,030	44,540
	Conventional	7,407	46,090
2nd (May 15)	No-tillage	6,403	42,370
	Conventional	6,590	43,870
3rd (June 1)	No-tillage	6,654	47,450
	Conventional	6,026	51,540

[1] Average of two years. Too wet to plant on this date in 1979.

probably need to be delayed to adjust for cooler soil temperatures (see Figure 2-6 for effect of planting date on corn yield), a special effort made to obtain adequate plant populations and more precise and careful management of nitrogen fertilizer will have to be practiced to avoid denitrification and leaching losses.

In latitudes where warming of the soil is slow and especially for the soils that have internal drainage problems, this wetness further slows the warming, and the addition of surface mulches and no-tillage adversely influence soil temperature. Under these soils and climatic conditions no-tillage may result in irregular germination, poor root development and slower growth resulting from sub-optimal soil temperature. In contrast, Lal (1976) concluded that the lower soil temperature associated with no-tillage management was a definite advantage in tropical areas.

SUITABILITY FOR SLOPING LAND

The adoption of no-tillage farming practices for row crop production has been heralded as the greatest soil conservation practice adopted during this century. For this reason no-tillage is especially well

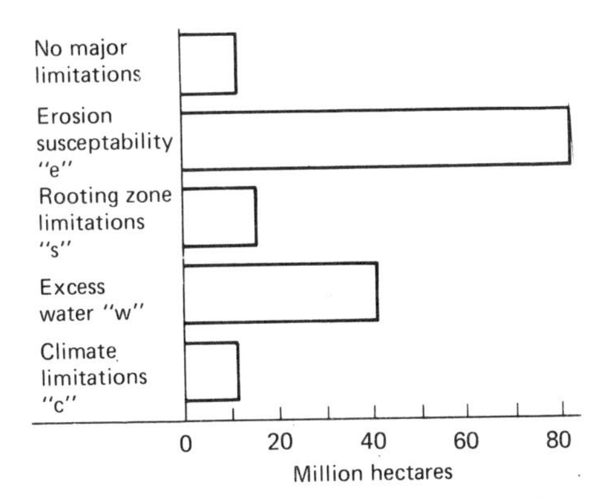

Figure 3-2. Comparison of soil limitations on cropland by subclass (from USDA-SCS, 1978, *The 1977 National Resources Inventory*).

suited for use on sloping lands. A comparison of soil limitations on cropland in the U.S. is presented in Figure 3-2. This documents the magnitude of soil erosion as a hazard and the problem is certainly one of major public concern. The intensity of land use on sloping land should be determined by an assessment of past soil erosion losses as well as considering the potential erosion hazard if these soils are tilled in the future.

The degree of erosion hazard is affected by length of slope, slope gradient, soil properties such as texture, structure, organic matter, along with rainfall intensity and cropping and tillage systems. Because of the gradient alone, sloping soils are inherently well to excessively well-drained. The effect of topography as a soil-forming factor has influenced the drainage characteristics of the sloping land. Potential surface runoff prevents excessive accumulation of water on these soils during rainy periods. As mentioned earlier in this chapter, the no-tillage systems of farming is best suited for well-drained soils. On wetter soils, the limiting problem may be removal of excessive water to allow the soil to dry and warm up enough for planting the crop. The no-tillage system effectively conserves soil and water.

Since land is a limited natural resource it is important to manage it prudently. As world population increases and food and fiber needs grow we can expect added pressures to use marginal land for crop production such as the potentially erosive sloping to steep lands of the world.

SOIL WATER EROSION

Soil erosion by water is usually a three-step process. These steps include: (1) detachment of soil particles, (2) transport of the soil and (3) deposition of sediment. Soil erosion by water occurs when conditions are favorable for detachment and transportation of soil particles. The amount of soil erosion that may occur is influenced by climate, soil erodibility, slope gradient and length of slope, and soil surface and vegetation conditions. Average annual precipitation (Figure 3-3) may influence the erosion hazard, however erosion losses are more closely related to the intensity of rainstorm and distribution of events. The actual water runoff has a greater influence on erosion. Water moving across the soil surface provides the mode of transportation and if soil particles are broken loose by raindrop impact or from the energy of the moving water these will be carried along as sediment. Average annual runoff in the U.S. is shown in Figure 3-4. Any management practice that can reduce raindrop impact on the soil, reduce water runoff and velocity of moving water, or improve the soil's resistance to erosion such as improving soil structure will reduce soil erosion losses. Erosion tolerances have been established for many soils of the world and these values can serve as guidelines as to how much soil erosion can be tolerated without losing their productive capacity over the long-term. The average annual sheet and rill erosion on croplands by states of the U.S. are shown in Figure 3-5. These losses on croplands usually exceed the allowable tolerances for much of the eastern section of the U.S. Although erosion losses are high in North America, areas of high-intensity rain storms and steep land such as Hawaii and many of the Caribbean islands often averaged greater than 60 mt/ha/yr losses (see Figure 3-6).

The Universal Soil Loss Equation (USLE) developed by Wischmeier and Smith (1965) can be used as a planning tool for developing soil management plans and selecting tillage systems. This equation can be used to predict soil water erosion losses for specific soils in a given climatic situation under different management systems.

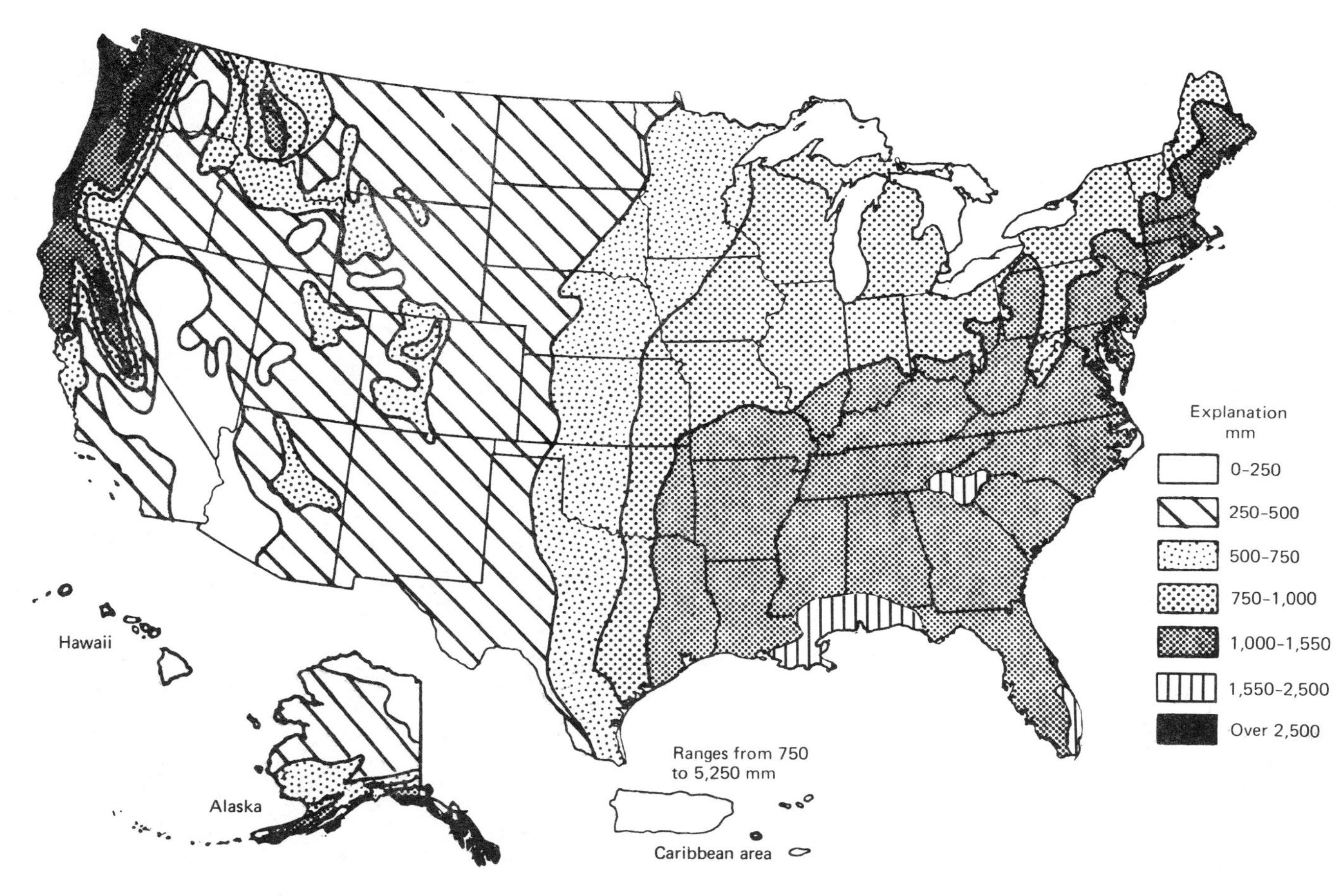

Figure 3-3. Average annual precipitation in the U.S. (from U.S. Water Resources Council, 1978).

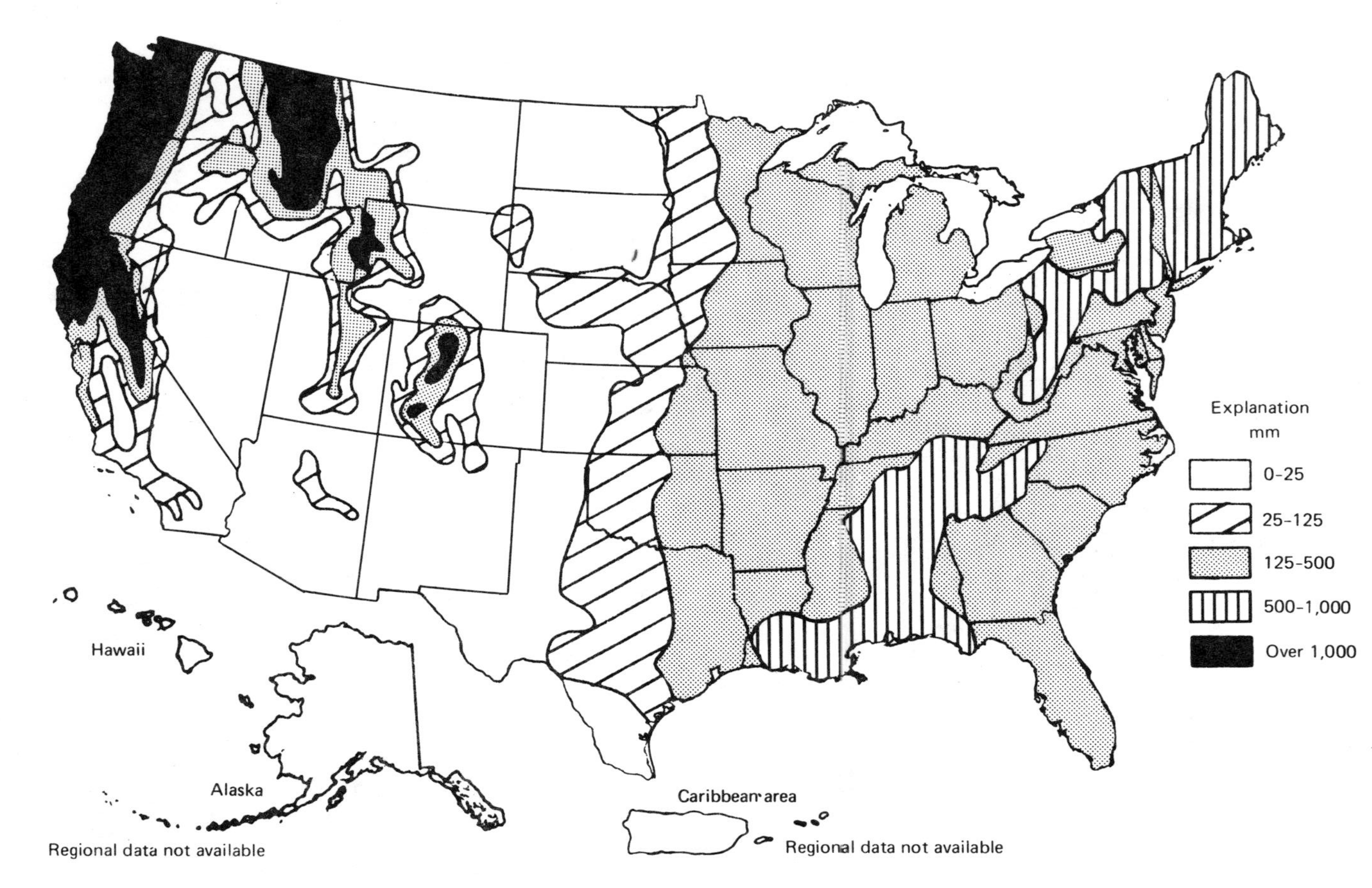

Figure 3-4. Average annual runoff in the U.S. (from U.S. Water Resources Council, 1978).

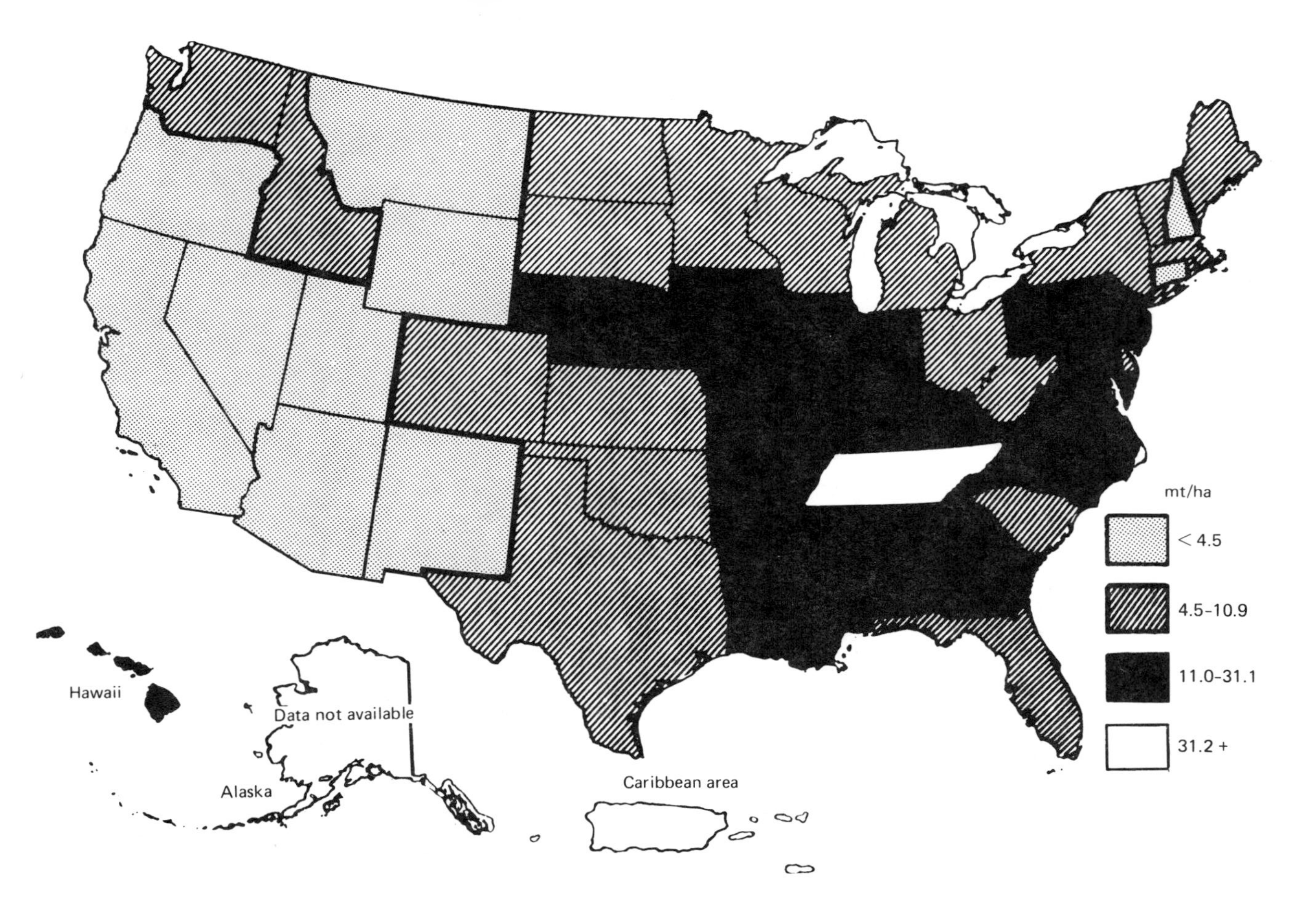

Figure 3-5. Average annual sheet and rill erosion on cropland (metric tons per hectare), by state (no data available for Alaska); the national average is 10.4 mt/ha/yr (from USDA-SCS, 1978, *The 1977 National Resources Inventory*).

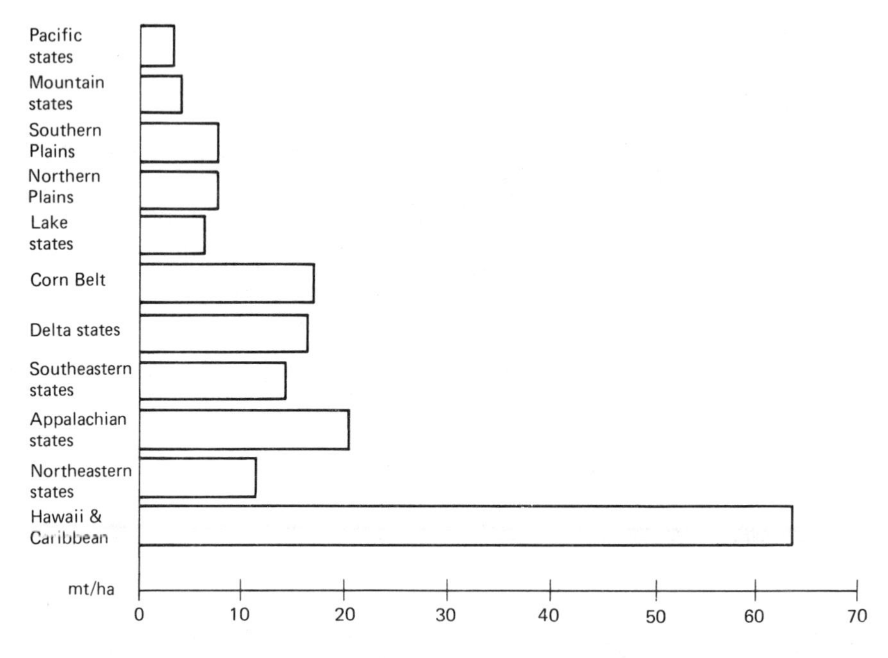

Figure 3-6. Estimated average annual sheet and rill erosion on cropland, by crop production region (from USDA-SCS, 1978, *The 1977 National Resources Inventory*).

The equation proposed by Wischmeier and Smith is:

$$A = R \times K \times LS \times C \times P$$

where

A = estimated average annual soil loss, mt/ha
R = rainfall erosion index
K = soil-erodibility factor, soil loss per unit of rainfall-erosivity index from bare fallow soil on a 9 percent slope, 22.1 m long
LS = slope length and steepness factor
C = cropping management factor
P = erosion-control-supporting-practice factor.

Although this equation can be useful for predicting erosion losses, it is often misused and misinterpreted. For example, not all the sediment lost from a slope leaves the field. It computes a long-term annual loss. Applying the equation to areas where the factors above are not well-known or defined may result in misuse. For example, well-tested and accurate C factors are not available for the various forms of no-tillage and reduced tillage.

EROSION CONTROL BY NO-TILLAGE

As no-tillage systems of farming evolve and are adopted, users agree that one of the major advantages of no-tillage is soil erosion control. Agencies concerned with conserving soil and water resources look to no-tillage techniques of row crop production as one of the best management practices available. This system allows more land to be utilized for food production and still keep soil losses to an allowable tolerance and prevent soil deterioration from excessive cultivations.

The effectiveness of no-tillage in controlling erosion is directly related to the amount of mulch maintained at the soil surface. Mulch rates required for satisfactory erosion control on sloping land have been investigated by Meyer *et al.* (1970), on sloping soils where erosion hazard is the major limitation for land use. The relationship of mulch to erosion losses by water in mt/ha are given in Table 3-4. It is interesting to note that even on slopes of 15 percent, relatively small levels of mulch produce marked reductions in erosion losses. A mulch rate of 4.5 tons uniformly distributed over the soil surface produced a 25-fold decrease in soil erosion losses. However it is important to understand that percent cover of soil surface is more

Table 3-4. Effect of straw mulch on soil erosion on a 15 percent slope. (From Meyer, Wischmeier, and Foster, 1970.)

MULCH (mt/ha)	EROSION LOSS (mt/ha)
0	62.3
1.12	19.4
2.24	11.5
4.48	2.5
8.96	1.5

important than the amount of surface mulch. For this reason, in some no-tillage systems it may be useful to shred or chop crop residues, such as corn stover, to acquire a more uniform cover over the soil surface. This is especially important if this is the only source of surface mulch in the cropping rotation.

No-tillage systems of production strongly influence the first stage of soil water erosion. First, the stable soil aggregates are not pulverized and destroyed by excessive tillage operations involved in seedbed preparations. Second, and probably more important, residues maintained on the soil surface are very effective in dissipating the energy from raindrop impacts during storms; this prevents the soil particles from being physically detached from the natural soil aggregates. Even though some overland flow of water may occur on soils under no-tillage management, movement of finer soil and organic fractions are often redeposited before the water and sediment move out of the area. The residue acts as a filter; therefore, soil water runoff from no-tillage plots do not carry a heavy load of sediments. Table 3-5 shows data from a severe rainstorm. Your attention is called to the fact that under no-tillage, a corn field with mulch at the surface and contour rows, about one-half of the total precipitation was lost as runoff, 6.4 cm, but only 0.07 mt/ha of soil was removed from the study area. On plowed rotation cornland with corn planted on the contour, runoff was almost equal to no-tillage, 5.8 cm, but soil loss was 7.1 mt/ha.

Another advantage from residue mulches maintained at soil surface is keeping the soil surface wetter and eliminating the problem associated with drying and crusting of the soil surface. Poorly

Table 3-5. Runoff and sediment loss from corn watershed, July 5, 1969. (From Harrold and Edwards, 1972, copyrighted by the Soil Conservation Society of America.)

TILLAGE	SLOPE (%)	RAINFALL (cm)	RUNOFF (cm)	EROSION (mt/ha)
Rotation plowed cornland, straight rows	7	14.0	11.2	49.8
Rotation plowed cornland, contour rows	6	14.0	5.8	7.1
No-tillage mulch cornland, contour rows	21	12.9	6.4	0.07

structured soils that characteristically develop due to sealing during rainfall occurrence and subsequently crusting upon drying can maintain a higher infiltration capacity under the no-tillage system (Lal, 1976; Triplett *et al.*, 1968). So, by reducing soil detachment by raindrop impact and flowing water and reducing water runoff by preventing crusting through use of surface mulches, the net effect is less soil loss by erosion.

These principles are well-documented by research conducted on soil erosion losses under different tillage and cropping systems. Research studies by Harrold and Edwards (1972) at Coshocton, Ohio on a Muskingum silt loam soil that is medium textured and well-drained highlights the differences in soil erosion associated with tillage systems. The data shown in Table 3-5 are associated with tests from a severe rainstorm but do reflect the relative differences in erosion under different management systems and conservation practices. The loss of 50 mt/ha of soil from a soil with a 6 percent slope, plowed and planted to corn in straight rows, demonstrates the magnitude of the erosion problem and why we should be concerned about soil erosion. The effectiveness of planting no-tillage corn, mulch left at the surface, and planted on the contour, is overly dramatized by the loss of 0.07 tons of soil. Also, this system was on a slope of 21 percent.

Harrold *et al.* (1967) reported measurements from watersheds for years where no severe rainstorms occurred and soil losses were low for all tillage systems. For example, in 1964 conventionally tilled treatments lost 6.3 mt/ha and no-tillage 0.1 mt/ha. During 1965 and 1966 soil erosion losses were not a factor on conventionally tilled or no-tillage plots (See Table 3-6).

Table 3-6. Soil loss and crop yields from conventional and no-tillage watersheds for three years at Coshocton, Ohio. (From Harrold *et al.*, 1967.)

	SOIL LOSS (mt/ha)		YIELD (kg/ha)	
YEAR	CONVENTIONAL	NO-TILLAGE	CONVENTIONAL	NO-TILLAGE
1964	6.3	0.1	5963	8537
1965	0.1	0	6654	6654
1966	0	0	6088	7344

Wittmus *et al.* (1973) suggested that by manipulation of the crop management factor "C value" of the soil loss equation, soil erosion can be reduced 95 percent. The soil loss equation considers the degree and frequency of tillage operations, residue cover and crop cover. Therefore, the ultimate in row crop production includes continuous no-tillage in a system that maintain at least 6–7 mt/ha of mulch at the surface as long as soil erosion is a concern.

The effectiveness of using mulches on the soil surface is presented in Table 3-7, which summarizes experimental studies conducted in the cornbelt area of the U.S. A combination of contouring and maintenance of residues at soil surface appear to be extremely effective in keeping soil erosion losses at acceptable tolerance levels.

Erosion studies in other areas have given similar results. McGregor *et al.* (1975) in North Mississippi reported that erosion was reduced from 17.5 mt/ha on conventionally tilled soybeans to 1.8 t/ha of soil loss under no-tillage management systems (Table 3-8). Studies on a claypan soil in Missouri (Kaniuka, 1975) compared water erosion losses of 9.6 mt/ha for conventionally tilled corn to 1.3 mt/ha for no-tillage corn. Also the no-tillage corn produced a higher grain yield.

Table 3-7. Effect of mulch and tillage system on soil losses at different locations in the cornbelt of the U.S. (From Mannering and Burwell, 1968.)

LOCATION, SOIL AND SLOPE	EXPERIMENTAL CONDITION	TILLAGE PRACTICE	SOIL LOSS (mt/ha)
Madison, Wis.	Simulated rainfall;	Conventional	50.0
Miami silt loam,	non-contoured	mulch	15.0
6% slope			
Madison, Wis.	Natural rainfall;	Conventional	3.2
Miami silt loam,	contoured	mulch	0.02
9% slope			
LaCrosse, Wis.	Natural rainfall;	Conventional	4.5
Fayette silt loam,	contoured	mulch	0.07
16% slope			
Coshocton, Ohio	Natural rainfall;	Conventional	17.5
Muckingrass silt loam,	contoured watershed	mulch	0.07
9–15% slopes	plots		
Indiana	Simulated rainfall;	Minimum	23.9
Russell silt loam,	non-contoured	mulch, chopped	1.1
5% slope		hay	

Table 3-8. Effect of tillage system on runoff and soil loss at Biloxi, Mississippi, three-year average. (From McGregor *et al.*, 1975.)

TILLAGE AND CROP SYSTEM	RUNOFF (%)	SOIL LOSS (mt/ha/yr)
No-till, soybean-wheat double crop	23	1.8
No-till, corn after soybeans	33	5.2
No-till, soybean after corn	24	1.3[1]
No-till, continuous soybeans	23	2.5
Conventional tillage, continuous soybeans	29	17.5

[1] Only two years' results.

Studies by Langdale *et al.* (1978) in Georgia on 6 percent sloping land showed that no-tillage plots with a rye cover crop as killed surface residue and planted to soybeans without tillage reduced soil losses from approximately 45 mt/ha to about 0.1 mt/ha.

CONTROLLING SOIL WIND EROSION BY NO-TILLAGE

The mechanics of wind erosion are similar to water erosion except wind becomes the transporting agent. Wind erosion is caused by wind blowing over an unprotected surface. After soil particles start to move, they are carried by the wind in three types of movement; suspension, saltation and surface creep. Once soil particles begin to move, they abrade the soil surface and break down soil aggregates. Like water erosion, wind erosion causes physical damage to growing crops, loss of soil fertility, loss of soil material, and pollution. The surface texture of soils may be altered by wind erosion through the removal of the finer particles moved by the wind, leaving behind a coarser fraction or sandier soil. During this removal of the finer soil fraction, the organic and clay fractions are lost and these are the focus of chemical activity and potential fertility components of the soil. The product left behind is often a sandy soil with a low-level productivity.

Since climatic factors related to wind erosion such as high temperature, low rainfall and high winds cannot be easily modified, the control of wind erosion must be through soil surface and crop residue management. Wind effects on soil in arid regions can be reduced by using management methods that conserve rainfall and reduce evaporation and by using tillage systems that do not pulverize the soil and destroy soil structure.

Stubble mulch farming is widely used in the western sector of the U.S. as a management for soil and crops and is very effective in reducing wind erosion. The no-tillage system can be used in conjunction with stubble mulching; one should improve this system by the use of no-tillage planters and chemical weed control so that disking operations can be omitted. The chisel plow that disrupts the soil below the soil surface, leaves the surface rough and allows 60–75 percent of the residues to remain at the soil surface is a very popular tillage tool and effective for conservation against wind erosion. A rough surface reduces the velocity of wind across the soil surface. Residues left on the soil surface help in maintaining a moist soil surface and serve as a barrier to the dislodging of finer particles from the soil aggregates. Because sandy soils are usually poorly aggregated, they are more susceptible to wind erosion than soils with high clay content and more stable soil aggregates.

Wind erosion experiments conducted by Schmidt and Triplett (1967), showed 291 mt/ha of material removed on land plowed and planted to corn as compared to a loss of 4.5 mt/ha from a no-tillage corn field during one severe wind storm.

In North America, soil erosion by wind is a major problem and concern in an area that extends from the Mississippi River to the Rocky Mountains and from the Gulf of Mexico to the prairies of Canada. Other geographic areas of concern include the muck and sandy areas of the Great Lakes Region and the sandy soils of the Gulf and Atlantic seaboards.

Conditions that facilitate wind erosion are poorly aggregated and bare soils, surface soils that are loose and dry and winds that are strong. Soil is especially susceptible to wind erosion when there is little vegetation to protect it. Some researchers feel the absence of vegetation is the primary cause of major losses due to wind erosion. Vegetative cover, according to Chepil (1957), is nature's way of protecting the earth's surface from erosion, and Chepil suggests that man has not been able to devise a better way. However, man's decisions related to the use and management of the land do have a direct

Table 3-9. Minimum amounts of standing stubble required to control wind erosion. (From Chepil *et al.*, 1961.)

TYPE OF RESIDUE	MINIMUM AMOUNTS REQUIRED (kg/ha)		
	FINE AND MEDIUM TEXTURED SOILS	SANDY LOAM SOILS	LOAMY SAND SOILS
Standing wheat (30 cm) stubble and growing wheat	840	1,400	1,960
Flattened wheat (30 cm) stubble	1,680	1,400	3,925
Grain sorghum stubble	1,680	2,800	3,925

influence on vegetative cover. Frequent cultivation of row crops such as cotton, beans and tobacco for the purpose of controlling weeds increase the erosion hazards because these cultural practices provide little cover to protect the soil. The erodibility of soil by wind is also affected by the size and strength of the soil structural units. This is another reason that stubble mulching and using a chisel plow is almost as effective as a completely no-tillage system.

Studies at Kansas State University, as reported by Chepil *et al.* (1961), provide data on the minimum amount of standing stubble required to control wind erosion, Table 3-9. These data also give an indication of how soil texture influences wind erosion. The coarser-textured loamy sands that have a single grain structure are more erosive than the medium- and fine-textured soils.

Prevention of wind erosion can be accomplished by windbreaks, crop residues, cropping and land use systems and tillage methods. The no-tillage system of crop production has a lot of potential in these areas where wind erosion is a potential problem. The features of no-tillage favor erosion control because of the very limited soil disturbance and because the maintenance of residues at soil surface reduces the surface wind velocity and serves as barrier to prevent the soil particles from being detached by the wind currents. Another advantage relates to conserving soil moisture; the surface is usually more moist than a bare cultivated soil.

STEEPLANDS

World population pressure is promoting the expansion of agriculture onto areas of marginal land such as steeper slopes and poorer soils. In many areas of tropical America, Asia and Africa, the better soils are often used for production of high-return export crops and more of the steeplands are being used for staple food production. Because of socioeconomic inequalities, many of the inhabitants of the steeplands areas have little alternative except to exploit these lands through crop production.

Agriculture in the steep mountains of many countries in Latin America make significant contributions to the national economy as well as providing staple foods to feed a large sector of the population. However, soil erosion has become a major threat to the continued productivity level for these soils. Although the highly skilled, soil and water conservation minded Incas of many years ago farmed these mountains successfully, the present inhabitants do not reflect such skills and technology. In Peru and Ecuador, for example, the highlands are extensively farmed to grain and vegetable crops with little or no attempt to control erosion. (See Figure 3-7).

On the extremely steep mountain sides, nothing short of permanent cover such as trees or grasses can reduce erosion losses to a reasonable level. Combinations of grassed contour ditches or terraces along with a reduced tillage system for crop production would at least reduce erosion losses many-fold.

In regions with large areas of steeplands, such as Nepal, India, Southeast Asia and Central and South America, soil erosion losses of >200 mt/ha are not uncommon. The low-lying prime farmland is also affected due to sedimentation deposits in the valleys and the siltation of streams and other water resources of the area.

The steeplands are usually inhabited by small land-owners with low per capita incomes. Thus, availability of capital to purchase needed production inputs is a severe limiting factor for crop production. This same scarcity of capital resources discourages the use of any soil conservation measure that requires a capital outlay.

Although research on suitability of no-tillage farming systems for these areas is very limited at this time, there is reason to believe that it could be adapted. It will be necessary to develop and test the proper combination of management inputs that are suitable for the diverse soil

Figure 3-7. Intensive farming on steeplands in Ecuador. Grain production by small farmers on long, steep slopes produces tremendous soil erosion losses.

and climatic variations of the steeplands. The use of backpack sprayers to apply herbicides to control weeds, and insecticides, if needed, will allow the small-hill farmers to produce more crops with the same amount of labor. Presently, crop production per farm unit is limited by the amount of land he can prepare for planting and subsequently control weeds by hand tools. No-tillage may reduce hand labor inputs by as much as 80–90 percent for certain crops, such as corn and soybeans.

There are many questions to be considered before deciding if no-tillage is a viable alternative cropping system suitable for the steeplands. These include the cost of herbicides; whether weeds can be effectively controlled; potential managerial skills of the farmers; whether disease and insect damage will be a major problem; development and availability of small, inexpensive equipment to plant on steep hillsides; and what the attitudes of the farmers are toward the adoption of new technology in crop production.

REFERENCES

Blevins, R. L., D. Cook, S. H. Phillips, and R. E. Phillips. 1971. Influence of no-tillage on soil moisture. *Agron. J.* **63**:593–596.

Bone, S. W., N. Rask, D. L. Forster, and B. W. Schurle. 1976. Evaluation of tillage systems for corn and soybeans. *Ohio Report, OARDC,* Wooster, Ohio, (July-August), pp. 60–63.

Cannell, R. Q., D. B. Davies, D. Mackney, and J. B. Pidgeon. 1978. The suitability of soils for sequential direct drilling of combine-harvested crops in Britain: A provisional classification. *Outlook on Agriculture* **9**:306–316.

Chepil, W. S. 1957. Erosion of soil by wind. In *Soil, Yearbook of Agriculture 1957.* U.S. Dept. Agric., Washington, D.C., U.S. Government Printing Office, pp. 308–314.

Chepil, W. S., N. P. Woodruff, and F. H. Siddoway. 1961. How to control soil blowing. *USDA, Farmers Bull. No. 2169* (September 7).

Gallaher, R. M. 1977. Soil moisture conservation and yield of crops no-till planted in rye. *Soil Sci. Soc. Am. J.* **41**:145–147.

Griffith, D. R., J. V. Mannering, H. M. Galloway, S. D. Parsons, and C. B. Richey. 1973. Effect of eight tillage-planting systems on soil temperature, percent stand, plant growth and yield of corn on five Indiana soils. *Agron. J.* **65**:321–326.

Harrold, L. L. and W. M. Edwards. 1972. A severe rainstorm test of no-till corn. *J. of Soil and Water Cons.* **27**:30. (Copyrighted by Soil Conservation Society of America.)

Harrold, L. L., G. B. Triplett, Jr., and R. E. Youker. 1967. Watershed tests of no-tillage corn. *J. Soil and Water Cons.* **22**:98–100.

Kaniuka, R. P., Editor. 1975. No-till systems prove ideal on claypan soils. *Agric. Res.* **24**:14 *(No. 1).*

Lal, R. 1976. No-tillage effects on soil properties under different crops in Western Nigeria. *Soil Sci. Soc. Am. J.* **40**:762–768.

Langdale, G. W., A. P. Barnett, and J. E. Box, Jr. 1978. Conservation tillage systems and their control of water erosion in the Southern Piedmont. *Proceedings of the First Annual Southeastern No-Tillage Systems Conference.* J. T. Touchton and D. G. Cummins, Editors. (Georgia Exp. Sta., Spec. Publication No. 5, Experiment, 1978), pp. 20–29.

Lessiter, Frank, Editor. 1982. 1981–1982 No-till farmer acreage survey. *No-Till Farmer.* **10**:5 (*No. 3,* March).

McGregor, K. C., J. D. Greer, and G. E. Gurley. 1975. Erosion control with no-tillage cropping practice. *Trans. Am. Soc. Agric. Eng.* **18**:918–920.

Meyer, L. D., W. K. Wischmeier, and G. R. Foster. 1970. Mulch rate required for erosion control on steep slopes. *Soil Sci. Soc. Am. Proc.* **34**:928–931.

Mannering, J. V. and R. E. Burwell. 1968. Tillage methods to reduce runoff and erosion in the corn belt. *Agr. Inf. Bull.* **330.**

Mitchell, W. H. and M. R. Teel. 1977. Winter annual cover crops for no-tillage corn production. *Agron. J.* **69**:569–573.

Moschler, W. W., D. C. Martens, C. I. Rich, and G. M. Shear. 1973. Comparative lime effects on continuous no-tillage and conventionally tilled corn. *Agron. J.* **65**:781–783.

Murdock, L. W. 1974. No-tillage on soils with restricted drainage. In *Proceedings of the No-Tillage Research Conference.* R. E. Phillips, Editor. University of Kentucky, Lexington.

Murdock, L. W. and R. L. Blevins. 1981. Unpublished data. Professor of Agronomy, University of Kentucky.

Schmidt, B. L. and G. B. Triplett, Jr. 1967. Controlling wind erosion. *Ohio Agric. Res. and Dev. Center Report* **52**:35–37.

Triplett, G. B. Jr., D. M. Van Doren, Jr., and B. L. Schmidt. 1968. Effect of corn (*Zea mays* L.) stover mulch on no-tillage corn yield and water infiltration. *Agron. J.* **60**:236–239.

United States Department of Agriculture, Soil Conservation Service. 1978. *1977 National Resources Inventories.*

United States Water Resources Council. 1978. The nation's water resources 1975–2000. *Second Natl. Water Assess.*

Van Doren, D. M. Jr., G. B. Triplett, Jr., and J. E. Henry. 1976. Influence of long term tillage, crop rotation, and soil type combinations on corn yield. *Soil Sci. Soc. Am. J.* **40**:100–105.

Wittmus, H. D., G. B. Triplett, Jr., and B. W. Greb. 1973. Concepts of conservation tillage systems using surface mulches. In *Conservation Tillage, Proceedings of a National Conference,* Soil Cons. Soc. Am., Ankeny, Iowa.

Wischmeier, W. H. and D. D. Smith. 1965. Predicting rainfall-erosion losses from cropland east of the Rocky Mountains. *Agriculture Handbook No. 282,* ARS-USDA, in cooperation with Purdue Agricultural Experiment Station.

4
Soil Moisture

Ronald E. Phillips
Professor of Agronomy
University of Kentucky

One advantage of no-tillage in row crop production is the conservation of soil water. The conservation of soil water is due to reduced soil water evaporation under no-tillage as compared to conventional tillage. This increased soil water content in no-tilled soil is available for transpiration through plants, thus increasing the water-use efficiency of the crop. No-tillage also significantly increases infiltration of water and reduces soil erosion. These topics are discussed in other chapters.

EFFECT OF MULCH ON EVAPORATION OF SOIL WATER

The no-tillage system of crop production implies that a mulch, usually consisting of plant residues, is present on the soil surface during the growing season or longer periods of time. Mulch as defined here is any material on the soil surface in which continuous liquid water films from the soil through the mulch are not present. Therefore, a mulch could be plant residues, gravel, sand, perlite, dry soil itself or any number of other materials. A dry layer of soil at the soil surface acts as a mulch; thus, the term self-mulching soils. Some sandy and clay soils are self-mulching.

The rate of loss of water or flux of water vapor through mulches is generally very slow in comparison to the rate of loss of water from a moist soil surface. The reasons for this are that: (1) in order for water to be lost by evaporation from a mulched soil, the water must change from a liquid to a vapor at the soil surface. The water vapor must then diffuse through the thickness of mulch which significantly reduces the rate of loss when compared to a bare soil surface; (2) the mulch reduces the quantity of direct solar radiation reaching the soil surface; thereby reducing the amount of energy available for change of state of water from liquid to vapor; and (3) mulches act as insulators to downward conduction of heat into the soil. The first two reasons given above cause, by far, the most reduction in loss of water by evaporation. Russel (1939) and Duley and Russel (1939) were the first researchers in the U.S. to apply the above principles to the conservation of water in dryland agriculture.

Reasons (2) and (3) are self-explanatory. To expand upon the first reason given above, the movement of water vapor results from diffusion, which is driven by a water vapor concentration or partial pressure gradient. Movement in the liquid phase is driven by a combination of gravitational and matrix potential gradients. Partial pressure gradients of water vapor in soil are not very great; for example, the relative humidity of air in soil at the wilting point (-15 bars) is approximately 98.9 percent. Considerable energy is required to change water from a liquid to a vapor; about 582 calories are required to change 1 g of liquid water to water vapor at a temperature of 25°C. This change in state of water often must take place several cm below the soil or mulch surface where the amount of energy available is much less than at the soil or mulch surface. The effect of depth of mulch on the relative flux of water vapor through the mulch via diffusion is shown in Figure 4-1. In Figure 4-1, the flux of water vapor through a mulch 0.1 cm thick is taken to be 100 percent; this approximates a bare, smooth, moist soil. A mulch 2 cm thick is about as effective in preventing soil water evaporation as a mulch 5 cm thick. For this reason, a dry layer of soil at the soil surface is very effective in reducing soil water evaporation as compared to a soil that is moist at the soil surface.

The flux, or quantity of water per unit area per unit time, of liquid water that can be conducted through soil to the soil surface from the soil below is described by Darcy's law, equation [1].

$$F = K(\Theta)\, \Delta H / \Delta X \qquad [1]$$

where F = flux of liquid water in cm^3 per unit area (cm^2) per unit time

$K(\Theta)$ = saturated or usually unsaturated conductivity in cm/time and is a function of the soil water content

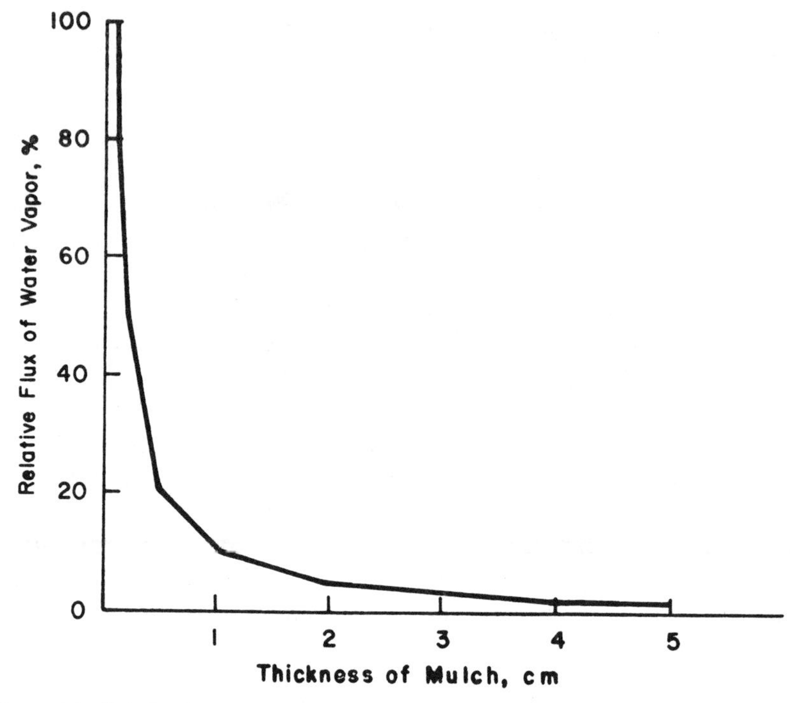

Figure 4-1. Relative flux of water vapor through a uniform mulch as a function of thickness of mulch; a mulch of 0.1 cm thick is taken to be 100 percent.

ΔH = the difference in the sum of the graviational potential and matrix potential in cm of water

ΔX = depth of soil in cm

In equation [1], the gravitation component of ΔH is usually very small in comparison to the matrix potential, which is always negative when the soil water content is less than saturated. The flux, F, from equation [1] is a product of two quantities $K(\Theta)$ and $\Delta H/\Delta X$.

Following a rain when the surface of a bare soil is moist, evaporation of water from the soil begins as soon as there is sufficient energy available to evaporate water. As evaporation proceeds, the water content of the soil decreases, and the unsaturated conductivity, K, also decreases. After some time, the amount of water being evaporated will exceed the flux of liquid water that can be conducted to the soil surface due to a decrease in K and usually a simultaneous decrease in $\Delta H/\Delta X$. When this happens, the surface soil begins to dry and will eventually self-mulch. The time required to form a dry layer of soil on the surface (self-mulch) is relatively longer when the evaporative potential at the soil surface is relatively low. Conversely, the time required to form a dry layer of soil on the surface is relatively short when the evaporative potential at the soil surface is relatively large. The time during which the rate of evaporation from an initially moist soil surface remains constant is called first stage evaporation. In the time following first stage evaporation, the rate of evaporation decreases. This is called the second stage of soil water evaporation. In these cases, it is assumed that evaporative potential at the soil surface remains constant. A schematic representation of this is shown in Figure 4-2. If the data in Figure 4-2 were plotted as cumulative evaporation versus time, then first state evaporation would be represented by the linear portions of the resulting curves. Curve *A* in Figure 4-2 could represent a bare soil, while curves *B*, *C* and *D* could represent the same soil with differing thicknesses of mulch; in addition, curves *A*, *B*, *C* and *D* could represent decreasing

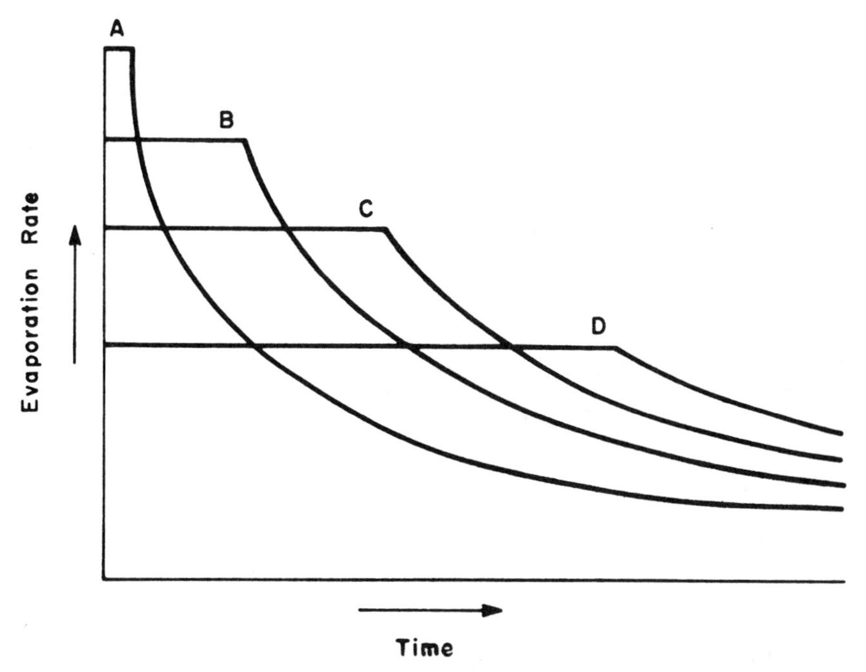

Figure 4-2. Schematic representation of soil water evaporation rate, as a function of time from soil(s) for differing conditions; see text.

evaporative potentials at the soil surface, respectively. Thus, it is seen that the presence of mulch on a soil surface will increase the time a soil remains in first stage evaporation but will decrease the rate of evaporation of soil water. A schematic representation of cumulative evaporation versus time for a bare soil and for the same soil with a mulch is shown in Figure 4-3. The mulched soil remains in first stage evaporation for the entire time shown in Figure 4-3, while the bare soil remains in first stage evaporation a relatively short time. If given sufficient time without rainfall, the cumulative evaporation of the mulched soil can exceed that of the bare soil because in the mulched soil, water will be lost from greater soil depths. Conservation of soil water will result in the mulched or no-tillage soil if rainfall occurs before the two curves cross each other. This time is approximately 30 days for Maury silt loam in Central Kentucky during the early part of the growing season (May and June). In order that the maximum amount of water be conserved due to a mulch or no-tillage, rainfall should occur at about time A (see Figure 4-3). When sufficient rainfall occurs to rewet the surface of the bare soil (Figure 4-3), first stage evaporation will again occur in the bare soil,

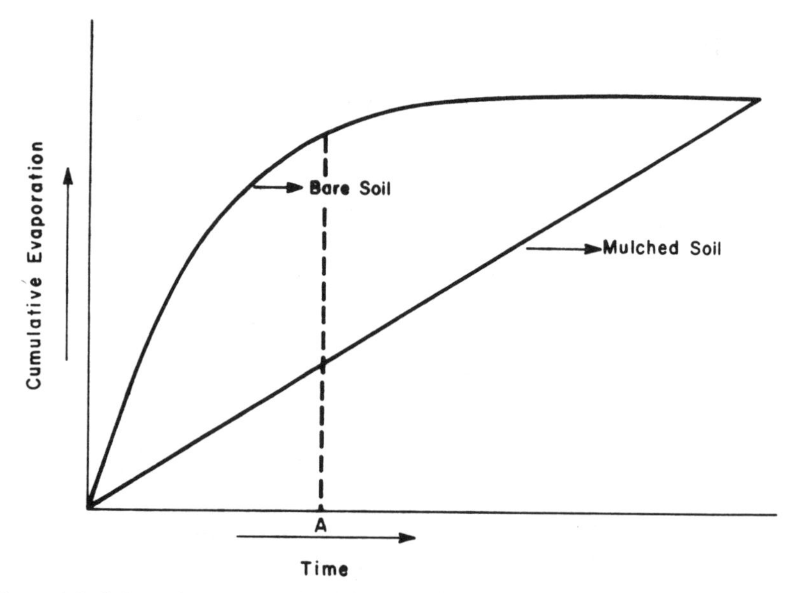

Figure 4-3. Schematic representation of cumulative evaporation of soil water as a function of time for a bare soil and the same soil with a mulch on the soil surface.

and the curve as shown in Figure 4-3 will be repeated. The more often this occurs, the more water is conserved due to no-tillage or mulch. This increased amount of water may then be transpired by a growing crop. The above principles of soil water evaporation apply in the early part of the growing season before a full crop canopy develops. After a full crop canopy develops, soil water evaporation is significantly reduced for the conventionally tilled soil. Therefore, most of the value of the conservation of soil water due to no-tillage or a mulch occurs before the development of a full crop canopy.

EFFECT OF SOIL TEXTURE, PANS AND DRAINAGE CLASS ON SOIL WATER EVAPORATION

Sandy soils and clay soils self-mulch more quickly than do medium-textured soils (silt loams). The reason for this is that the unsaturated conductivity, $K(\Theta)$ (see equation [1]), is too small to conduct liquid water to the soil surface under the normal evaporative potential existing during the growing season of agronomic crops. The unsaturated conductivity of sandy soils at field capacity is relatively large even though the water content of sandy soils at "field capacity" is relatively small. The saturated conductivity decreases rapidly as the water content of the surface decreases due to evaporation; K may decrease several orders of magnitude for a small change in the water content. A dry layer of the sandy soil develops very quickly as soon as the quantity of water conducted to the soil surface is less than the quantity of water evaporated. Thus, sandy soils are generally classified as self-mulching soils.

The unsaturated conductivity, K, of clay soils at "field capacity" is relatively small even though clay soils hold relatively large amounts of water at "field capacity." However, the unsaturated conductivity of clay soils decreases much more slowly than that of sandy soils as the water content decreases below "field capacity." Normally, the evaporative potential which exists during the period from planting time of agronomic crops until the plants begin to transpire significant amounts of water will result in a greater amount of water evaporation than can be conducted to the soil surface of the clay soil due to the low K of clay soils. A dry layer of soil quickly develops on the surface; thus, clay soils are generally classified as self-mulching. It should be pointed out that K of clay soils decreases relatively slowly as the water content decreases below "field capacity." For this reason, clay soils will remain in first stage evaporation a relatively long time if the evaporative potential is low when the air temperature is below 10°C on a clear sunny day but will remain in first stage evaporation a short time when the evaporative potential is high when the air temperature is above 20°C on a clear, sunny day.

The unsaturated conductivity of silt loam soils or medium-textured soils at "field capacity" is intermediate between those for sandy and clay soils. The decrease in K of silt loam soils as the water content decreases below "field capacity" is also intermediate between those for sandy and clay soils but is much closer to that for clay soils than that for sandy soils. Therefore, silt loam soils will remain in first stage evaporation for a longer period of time than will a sandy soil or a clay soil for the evaporative potential that exists at 20°C and higher on clear, sunny days.

As pointed out by Gardner (1958), evaporation of soil water is significantly influenced by the presence of a permanent or perched water table near the soil surface (the surface of a water table is defined to be the depth of the free water surface or where the water is at atmospheric pressure). There is often a zone of soil above the surface of the water table where the soil is saturated but the water is held under negative pressure. This zone is termed the capillary fringe. Some soils under certain situations can remain in first stage evaporation for long periods of time. This can occur if the water table is close enough to the soil surface such that the rate of upward movement of water from the water table is equal to the rate of evaporation. For such situations, a mulch or no-tillage is not particularly desirable because the mulch will cause the soil surface to remain wet for a much longer period of time than would be the case in the absence of a mulch. In such cases, the plow layer may become anaerobic.

Many soils of agricultural importance have pans within the plant rooting depth. Pans can be beneficial to a growing crop or they can be harmful as far as supplying water to a crop is concerned, depending upon the depth of the pan in the rooting zone and the amount and distribution of rainfall. The Zanesville, Grenada and Calloway soils in western Kentucky are such soils. The pan in Zanesville silt loam occurs at a depth of about 70 cm. Due to the presence of a perched water table above the pan, the Zanesville soil often remains very wet or even saturated in the plow layer during April and May, which means that corn often must be planted several weeks later than is the case in well-drained soils in the area. It is not desirable to have a mulch on these soils during April and May, since the plow layer will dry out more slowly with a mulch than without a mulch. The desirability of no-tillage on this soil during June and July depends upon the amount and distribution of rainfall during June and July. In general, in years of below average rainfall, the presence of a mulch on the soil surface during June and July will benefit a crop due to the conservation of soil water. In years of above-average rainfall, the presence of a mulch often causes poor aeration due to wetness in the rooting depth.

SOIL WATER IN FIELD STUDIES

The plant-extractable water for no-tillage and conventional tillage treatments in the surface 45 cm of Maury silt loam soil located in central Kentucky is shown in Figure 4-4. Rainfall amounts and distribution are also shown. Corn was planted on May 10 and harvested on September 14. Grain yields were 11,045 and 11,795 kg/ha for the conventional tillage and no-tillage treatments, respectively. As can be seen in Figure 4-4, plant-extractable water in the 0–45-cm depth of soil of the conventional tillage treatment decreased from 8.8 cm on May 14 to 0 cm on August 25, while plant-extractable water of the no-tillage treatment did not decrease appreciably until after July 26 and then decreased to a minimum on August 25. On August 25 there were 0 and 2.3 cm of plant extractable water in the conventional-tillage and no-tillage treatments, respectively.

Even though there was 7.4 cm of rain from May 10 to June 12, 9.4 cm from June 13 to June 21, 14.4 cm from June 22 to July 23 and 5.8 cm from July 24 to August 25, the soil water content decreased continuously from May 10 to August 25 on the conventional tillage treatment; there was a total of 37.0 cm of rainfall from May 10 to August 25. The soil water data presented in Figure 4-4 presented for no-tillage and conventional tillage are typical of such data on Maury silt loam.

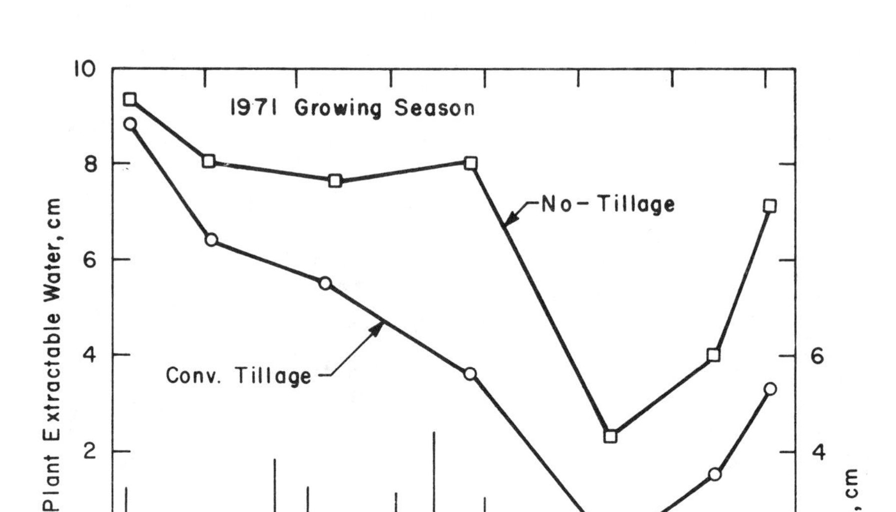

Fig. 4-4. Plant extractable water remaining in the surface 45 cm of Maury silt loam soil during the 1971 growing for corn grown under no-tillage and conventional-tillage; precipitation amounts and distribution are also shown.

The data presented in Figure 4-4 are from an experiment initiated in 1970 that has continued until the present time. The slope of experimental site is 0–2 percent. Runoff of water from rainstorms has never been observed to occur from either the no-tilled or conventionally tilled plots. The differences between soil water contents of the two treatments at lower soil depths than reported in Figure 4-4 have been considerably less than in the 0–45-cm depth, but the soil water content at the lower depths has always been as high or higher in the no-tillage treatment than in the conventionally-tilled treatment. The plant-extractable water in the 0–45-cm and 45–90-cm depths of the Maury silt loam are 8.8 and 4.8 cm, respectively. The rooting depth in the Maury soil is not much greater than 90 cm. Blevins and Cook (1970), Blevins *et al.* (1971), Na Nagara (1972) and Zartman (1974) have published data showing the effect of no-tillage on soil water in central Kentucky soils all of which are well-drained soils.

Estimated amounts of soil water evaporation and transpiration by corn from no-tilled and conventionally tilled treatments by month during the growing seasons of 1970 through 1973 are shown in Table 4-1. These estimates were calculated from the model proposed by Ritchie (1972) as modified by Radcliffe *et al.* (1980). The means of the four years are shown at the bottom of Table 4-1. Soil water evaporation was less in every month and every year in the no-tilled treatment (average of 4.5 cm) than in the conventionally tilled treatment (average of 15.0 cm). Therefore, more water was available for transpiration in the no-tilled plots than in the conventionally tilled plots; the average transpiration during the growing season was 30.7 and 24.2 cm for the no-till and conventionally tilled treatments, respectively. The corresponding evapotranspiration for the no-till and conventional treatments was 34.8 and 43.3 cm, respectively. The average difference in transpiration between the two treatments, 30.7 cm for the no-tilled treatment and 24.2 cm for the conventionally tilled treatment, can often result in the difference between an excellent yield of grain and a medium yield of grain, depending upon the stage of growth when the difference in transpiration occurs. The estimates of soil water evaporation and transpiration in Table 4-1 from the conventional tillage system are consistent with the data of Peters (1960) and Harrold *et al.* (1959). They estimated that evapotranspiration from corn in the north central U.S. was composed of approximately 50 percent evaporation and 50 percent transpiration.

Estimates of the amount of soil water adsorbed by roots and subsequently transpired by corn from several depths in the soil are given in Table 4-2. For the time periods given in Table 4-2, corn growing on the no-tillage plots transpired more water than corn growing on conventional tillage plots. This was due for the most part to more water taken from the 30–45-cm soil depth for the no-tillage treatment than for the conventional tillage treatment. These data again illustrate that the amount of soil water present in the rooting depth of no-tillage plots is as great or greater than the amount present in the rooting depth of conventionally tilled plots.

Table 4-1. Estimated soil water evaporation and transpiration by corn from no-tilled and conventionally-tilled treatments on Maury silt loam soil by month during the 1970, 1971, 1972 and 1973 growing seasons.

		NO-TILL		CONVENTIONAL TILL		RAINFALL (cm)
Year	Month	Transpiration (cm)	Evaporation (cm)	Transpiration (cm)	Evaporation (cm)	
1970[1]	May	0.1	3.0	0.0	5.4	19.9
	June	10.3	0.8	7.5	6.8	9.2
	July	10.8	0.2	7.1	1.5	5.2
	August	2.3	0.1	2.9	0.5	3.3
	September	1.3	0.7	0.7	3.1	12.0
	Total	24.8	4.8	18.2	17.3	49.6
1971[1]	May	0.0	1.6	0.0	6.2	17.7
	June	7.1	1.2	6.3	5.7	9.7
	July	12.9	0.3	11.5	2.4	17.9
	August	11.5	0.3	9.9	1.9	2.0
	September	2.3	0.8	2.0	5.5	13.1
	Total	33.8	4.2	29.7	21.7	60.4
1972[1]	May	0.0	2.7	0.0	5.1	8.8
	June	6.3	1.0	6.0	6.1	5.5
	July	11.6	0.3	7.5	1.9	6.5
	August	11.0	0.2	5.9	1.2	7.1
	September	1.3	0.3	1.1	0.7	7.3
	Total	30.2	4.5	20.5	15.0	35.2
1973[1]	May	0.0	1.2	0.0	8.5	15.1
	June	6.8	1.2	6.0	8.6	14.5
	July	14.2	0.3	11.9	2.6	10.7
	August	11.9	0.2	10.0	1.8	3.9
	September	1.0	0.3	0.5	0.7	3.9
	Total	33.9	3.2	28.4	22.2	48.1
Average of Four Years						
	May	0.0	2.1	0.0	6.3	17.9
	June	7.6	1.0	6.4	6.8	9.7
	July	12.4	0.3	9.5	2.1	10.1
	August	9.2	0.2	7.2	1.4	4.1
	September	1.5	0.5	1.1	2.5	9.1
	Total	30.7	4.1	24.2	19.1	50.9

[1] Planting dates in 1970, 1971, 1972 and 1973 were May 6, May 10, May 17 and mid-May, respectively.

Table 4-2. Estimates of the amount of soil water transpired by corn from several depths in the rooting zone for selected time periods in 1971 and 1972.

TREATMENT	SOIL DEPTH (cm)	WATER TRANSPIRED (cm)		
		28-53 DAYS AFTER PLANTING, 1971	54-76 DAYS AFTER PLANTING, 1971	49-97 DAYS AFTER PLANTING, 1972
Conventional tillage	0-15	3.8	3.9	6.5
	15-30	3.2	2.8	3.1
	30-45	1.2	0.7	1.0
	45-60	0.3	0.6	0.9
	60-75	0.2	0.4	0.7
	75-90	–	–	–
Total		8.7	8.4	12.9
cm/day		0.35	0.38	0.27
Rainfall, cm		9.65	15.0	10.4
No-tillage	0-15	3.4	3.2	7.6
	15-30	2.7	2.6	4.0
	30-45	1.9	1.9	1.5
	45-60	1.0	1.1	0.9
	60-75	0.1	0.1	0.5
	75-90	0.0	–	–
Total		9.1	8.9	14.5
cm/day		0.36	0.41	0.30
Rainfall, cm		9.65	15.0	10.4

Table 4-3 shows a comparison of water use efficiency of corn during several time periods during the 1971 growing season and for the entire 1970 and 1971 growing seasons for no-tillage and conventional tillage. The values given for no-tillage over-estimate the actual values because drainage of water below the rooting depth was included in the calculations as well as transpiration and soil water evaporation. We have observed that more drainage of soil water below the rooting depth occurs under no-tillage than occurs under conventional-tillage during most growing seasons. It was estimated that 1.0 and 4.8 cm of water drained below the rooting depth on the conventional tillage treatment in 1970 and 1971, respectively. The corresponding values for the no-tillage treatment were 2.5 and 10.0 cm in 1970 and 1971, respectively.

Table 4-3. Water use efficiency (grams of water required to produce one gram of dry matter during four periods of the 1971 growing season and for the entire 1970 and 1971 growing seasons). (From Na Nagara, 1972.)

	WATER TRANSPIRED (cm)			
	CONVENTIONAL TILLAGE		NO-TILLAGE	
TIME PERIOD	0 kg N/ha	341 kg N/ha	0 kg N/ha	341 kg N/ha
May 14–June 10, 1971	7,254	7,440	6,064	5,886
June 10–June 25, 1971	756	691	771	577
June 25–July 26, 1971	230	196	226	141
July 26–August 26, 1971	161	145	318[1]	166
May 14–August 26, 1971	299	258	364	215
May 18–August 18, 1970	235	195	222	186

[1] Nitrogen deficiency symptoms in corn were apparent.

During the early part of the growing season, evaporation of soil water accounted for almost all of the water use. As the corn grew larger, the soil surface became increasingly shaded, and evaporation of soil water was much less than transpiration of water through the corn plants.

The difference in soil water content of no-tillage as compared to conventional tillage has an influence upon nitrogen fertilizer practices and timing of application. For a given amount of rainfall, the rainwater will move deeper into the soil profile for no-tillage than for conventional tillage due to the higher soil water content of no-tillage than for conventional tillage. Tyler and Thomas (1977) in June 1973 measured a total of 14.7 kg/ha of nitrate nitrogen in leachate collected with a lysimeter for no-tillage in a Maury silt loam soil; only 4.0 kg/ha was found in leachate through the soil profile under conventional tillage. In another study, without the use of a lysimeter, McMahon and Thomas (1976) found appreciable leaching of nitrate and chloride during the 1971 growing season under no-tillage, while a lesser amount of leaching of nitrate or chloride occurred under conventional tillage. During the winter of 1971–72, leaching was similar on both tillage treatment, but because of the differences in leaching during the growing season, essentially all of the chloride and nitrate were removed from the 0–90-cm soil depth under no-tillage and only about half from the conventionally tilled soil by April 1972. No crop was grown on this experimental soil during 1971. The increase of soil water evaporation of conventional tillage as compared to no-tillage undoubtedly contributes to less nitrate nitrogen leaching through the soil profile in conventional tillage than in no-tillage.

In addition to the factors listed above, flow of rainwater down macropores also appears to be a factor in nitrate nitrogen, and other mobile cations and anions, leaching more under no-tillage than under conventional tillage. Wild (1972), Blake *et al.* (1973), McMahon and Thomas (1974), Ritchie *et al.* (1972), Thomas *et al.* (1973, 1978), Shuford *et al.* (1977) and Quisenberry and Phillips (1976, 1978) show that significant amounts of water movement can and do occur in soil macropores. Thomas and Phillips (1979) point out the consequences of water movement in macropores, including mobile anion and cation movement through soils. Moody *et al.* (1963) found that 6.7 mt/ha of chopped straw on these soil surfaces increased the plant-available water in the 0–46-cm soil depth on the average of 1.2, 2.9 and 2.2 cm of water during the growing seasons of 1958, 1959 and 1960 growing seasons, respectively. The grain yields at 84 kg of N/ha were 5,485, 3,553 and 3,675 kg/ha for the unmulched soil for the 1958, 1959 and 1960 growing seasons, respectively. The corresponding yields for the mulched soil were 7,356, 8,184 and 6,340 kg/ha. They attributed the higher yields to the greater amount of plant-available water in the mulched soil. In a later publication, Jones *et al.* (1969) evaluated the effect of no-tillage and conventional tillage corn with and without a mulch on the soil surface. They found that mulched treatments, whether of undisturbed tilled sod on the no-tillage plots or of straw on conventional tillage plots, gave the lowest values for runoff and the highest values for soil water content and yield of grain. Soil water conserved by the mulches resulted in an average grain yield increase of 1,932 kg/ha. The effect of tillage *per se* was minor. Triplett *et al.* (1968), in Ohio, on the other hand, found that no-tillage without a mulch significantly reduced corn grain yields (4,770 kg/ha) as compared to conventional tillage (5,970 kg/ha), but no-tillage with a mulch significantly increased yields (6,625 kg/ha). Both the data of Triplett *et al.* (1968) and Jones *et al.* (1969) point out the value of a mulch in the no-tillage system.

Nelson *et al.* (1977) reported soil water potential of conventional and no-tillage over a four year period from an experiment in Georgia on Cecil sandy clay loam soil; they did not report soil water contents. Soil water potentials, average of 15- and 40-cm depths, in conventional tillage, were equal to or lower than (more negative) no-tillage throughout the growing season. This study involved corn and sorghum in double cropping systems; both corn and sorghum grain yields were as high or higher on no-tillage systems as on conventional tillage systems.

Lal (1974) in the tropics (Western Nigeria) found significantly higher soil water content throughout the growing season in the 0–10-cm depth and for most of the growing season in the 10–20-cm soil depth for no-tillage than for conventional tillage. The soil water holding capacity as well as the soil water release characteristics were, also, higher for no-tillage. Agboola (1981) found similar results in the same general area.

Gantzer and Blake (1978) in Minnesota, on a clay loam soil, found that volumetric soil water content in the surface was in the range of 0.35–0.28 percent for no-tillage as compared to a range of 0.31–0.25 percent for conventional tillage.

Unger (1978) in Texas, where fallow is often practiced to increase soil water in the soil rooting depth for a row crop the following year, found that mulch increased available soil water at planting due to the fallowing period the previous year an average of 0.68 cm/mt mulch/ha. He attributes the increased plant-available water to decrease soil water evaporation and increased infiltration in the mulched treatments. Grain sorghum yields averaged 1,780 and 3,990 kg/ha for the bare and 12 mt/ha, respectively. Water-use efficiency increased from 55.6 kg/ha-cm for bare soil to 115.0 kg/ha-cm for 12 mt of mulch/ha. Mulch rates used were 0, 1, 2, 4, 8 and 12 mt/ha. A tilled soil with mulch behaves very similarly to no-tilled soil as far as soil water conservation is concerned if they have the same kind and amount of mulch on the soil surface. Small differences can result due to differences in soil physical properties.

Triplett *et al.* (1973) estimates that there are 4.2 million hectares in Ohio that are well suited for no-tillage; i.e., no-tillage corn production should be equal to or greater than any other tillage system. This value includes active and idle cropland, pasture and forest. Of this

4.2 million hectares, 2.3 million hectares are adequately-drained soil and cleared of trees and should be immediately available for no-tillage corn production for farmers desiring it. Soils grouped in this category are moderately well-, well- and excessively well-drained or shallow and they have silt loam, loam, sandy loam or loamy fine sand surface texture.

EFFECT OF NO-TILLAGE ON ROOT GROWTH

Because no-tillage or mulch greatly reduces soil water evaporation at the soil surface and consequently increases soil water content at the soil surface, it is reasonable to assume no-tillage or a mulch will affect plant root growth and distribution, especially in the first few surface centimeters. Data to substantiate this hypothesis are given in Table 4-4. Obviously, the corn root density in no-tillage at the 0–5-cm soil depth is much larger than under conventional tillage. Little difference existed between the two treatments below 5 cm. In addition, there was little difference in root density of the two treatments when averaged over the 0–30-cm soil depth. More complete corn root data for no-tillage and conventional tillage were reported by Na Nagara *et al.* (1976) and are given in Table 4-5. Instead of sampling small soil depth increments near the soil surface, Na Nagara *et al.* (1976) sampled to a soil depth of 90 cm. Root samples were

Table 4-4. Comparison of corn root density in Zanesville silt loam soil for corn grown under no-tillage and conventional tillage. Root samples were taken 15 cm from center of row. (From Phillips *et al.*, 1971.)

	CONVENTIONAL TILLAGE		NO-TILLAGE	
SOIL DEPTH	JUNE 10	JULY 6	JUNE 10	JULY 6
cm	(cm of root/cm^3 of soil)			
0–5	1.6	1.9	21.2	21.0
5–15	6.9	5.1	5.2	5.3
15–30	2.6	8.4	1.0	2.6
0–30	3.9	6.2	5.8	6.6

Table 4-5. Average root density by soil depth of corn growing on Maury silt loam soil during several time periods during the 1972 growing season for no-tillage, and conventional tillage. Root samples were taken 30 cm from the corn row. (From Na Nagara *et al.*, 1976.)

DAYS AFTER PLANTING	NITROGEN RATE (kg N/ha)	SOIL DEPTH (cm)	ROOT DENSITY NO-TILLAGE	ROOT DENSITY CONVENTIONAL TILLAGE
			(cm of root/cm³ of soil)	
34–49	0	0–15	1.08	0.58
		15–30	0.21	0.66
		30–45	0.08	0.24
		45–60	0.06	0.09
	168	0–15	0.98	0.30
		15–30	0.18	0.29
		30–45	0.30	0.12
		45–60	0.06	0.04
49–76	0	0–15	1.26	0.69
		15–30	0.32	0.68
		30–45	0.25	0.36
		45–60	0.12	0.35
		75–90	0.03	0.22
	168	0–15	0.77	0.54
		15–30	0.32	0.31
		30–45	0.59	0.37
		45–60	0.10	0.36
		75–90	0.00	0.10
79–97	0	0–15	2.27	0.92
		15–30	0.54	0.75
		30–45	0.45	0.33
		45–60	0.28	0.82
		60–75	0.35	0.35
		75–90	0.07	0.12
	168	0–15	0.50	0.71
		15–30	0.54	0.72
		30–45	0.60	0.37
		45–60	0.17	0.38
		60–75	0.33	0.17
		75–90	0.01	0.07

also taken at different distances from the row, 30 cm for Maury and 15 cm for Zanesville. The root density values reported in Table 4-5 for Maury silt loam soil are much smaller than those reported in Table 4-4 for Zanesville silt loam soil. The authors were not able to explain why plant roots do not elongate as profusely in Maury silt loam soil as they do in most other soils. In Table 4-5, as was the case for the data shown in Table 4-4, the corn root density in the 0–15-cm soil depth is greater for no-tillage than for conventional tillage, and the root density in the 15–30-cm soil depth is greater for conventional tillage than for no-tillage.

Zartman *et al.* (1976) reported root density of tobacco grown in Maury silt loam soil under no-tillage and conventional tillage. These data are shown in Table 4-6. The root density of tobacco under no-tillage was less than for conventional tillage during most of the growing season. This difference was probably due to both the structure of the soil of the two treatments at the time of transplanting and to the establishment of the bare-rooted transplants. A loose seedbed was prepared for transplanting in conventional tillage while no comparable seedbed was prepared for the no-tillage treatment. Apparently, the tobacco roots of the transplants were restricted to the transplanting trench under no-tillage, while a transplanting trench was not distinguishable in the conventional tillage treatment following transplanting. There are several other differences in the two treatments which could cause differences in growth of roots. First, tobacco is a bare-rooted, transplanted crop, while corn is a direct seeded crop. The soil-root contact of a tobacco plant transplanted into a killed sod (no-tillage) is probably not as good as when transplanted into a well-worked, conventionally tilled seedbed. Second, the soil temperature in the rooting zone of no-tillage is lower than the temperature in conventional tillage at transplanting time (Phillips, 1974). While this is true for both corn and tobacco, it may be more deleterious for tobacco than corn. Third, the microclimate just above the soil surface at transplanting is undoubtedly different for conventional-tillage and no-tillage. Increased temperature above the soil surface for the no-tillage treatment could cause leaf injury, resulting in a lower survival rate and slower growth rate when compared with the conventional tillage treatment.

Table 4-6. Root density of tobacco growing in Maury silt loam soil, 1972, as a function of soil depth and time for no-tillage and conventional tillage at two rates of nitrogen fertilization. Samples were taken a distance of 10 cm from the plant. (From Zartman *et al.*, 1976.)

DAYS AFTER TRANSPLANTING	NITROGEN RATE (kg N/ha)	SOIL DEPTH (cm)	ROOT DENSITY	
			NO-TILLAGE	CONVENTIONAL TILLAGE
			(cm of root/cm^3 of soil)	
35	90	0–15	0.21	0.81
		15–30	0.02	0.08
		30–45	0.00	0.02
	180	0–15	0.31	0.36
		15–30	0.00	0.09
		30–45	0.00	0.00
49	90	0–15	0.27	1.73
		15–30	0.10	0.35
		30–45	0.02	0.05
	180	0–15	0.56	1.02
		15–30	0.06	0.45
		30–45	0.03	0.06
63	90	0–15	0.62	1.43
		15–30	0.11	0.71
		30–45	0.13	0.53
		45–60	0.19	0.12
	180	0–15	1.19	1.57
		15–30	0.52	0.47
		30–45	0.32	0.55
		45–60	0.20	0.13
77	90	0–15	0.19	1.36
		15–30	0.12	0.42
		30–45	0.05	0.46
		45–60	0.00	0.14
	180	0–15	0.23	0.74
		15–30	0.14	0.23
		30–45	0.06	0.14
		45–60	0.00	0.08

REFERENCES

Agboola, Akinola A. 1981. The effects of different soil tillage and management practices on the physical and chemical properties of soil and maize yield in a rainforest zone of Western Nigeria. *Agron. J.* **73**:247–251.

Blake, G., E. Schlichting, and U. Zimmerman. 1973. Water recharge in a soil with shrinkage cracks. *Soil Sci. Soc. Am. Proc.* **37**:669–672.

Blevins, R. L. and D. Cook. 1970. No-tillage: Its influence on soil moisture and temperature. 187. *Agr. Exp. Sta. Progress Report* **187**, Lexington, Kentucky.

Blevins, R. L., D. Cook, S. H. Phillips, and R. E. Phillips. 1971. Influence of no-tillage on soil moisture. *Agron. J.* **63**:593–596.

Duley, F. L. and J. C. Russel, 1939. The use of crop residues for soil and moisture conservation. *J. Am. Soc. Agron.* **3**:703–709.

Gantzer, C. J. and G. R. Blake. 1978. Physical characteristics of LeSuer clay loam soil following no-till and conventional tillage. *Agron. J.* **70**:853–857.

Gardner, W. R. 1958. Some steady state solutions of the unsaturated moisture flow equation with application to evaporation from a water table. *Soil Sci.* **85**:228–232.

Harrold, L. L., D. B. Peters, F. R. Dreibelbis, and J. L. McGuiness. 1959. Transpiration of corn grown on a plastic covered lysimeter. *Soil Sci. Soc. Am. Proc.* **23**:174–178.

Jones, J. N., J. E. Moody, and J. H. Lillard. 1969. Effects of tillage, no-tillage and mulch on soil water and plant growth. *Agron. J.* **61**:719–721.

Jones, J. M., J. E. Moody, G. M. Shear, W. W. Moschler, and J. H. Lillard. 1968. The no-tillage system for corn. *Agron. J.* **60**:17–20.

Lal, Rattan. 1974. No-tillage effects on soil properties and maize (*Zea mays* L.) production in Western Nigeria. *Pl. and Soil* **40**:321–331.

McMahon, M. and G. W. Thomas. 1974. Chloride and triatiated water flow in disturbed and undisturbed soil cores. *Soil Sci. Soc. Am. Proc.* **38**:727–732.

McMahon, M. A. and G. W. Thomas. 1976. Anion leaching in two Kentucky soils under conventional-tillage and a killed-sod mulch. *Agron. J.* **68**:437–442.

Moody, J. E., J. N. Jones, and J. H. Lillard. 1963. Influence of straw mulch on soil moisture, soil temperature and growth of corn. *Soil Sci Soc. Am. Proc.* **27**:700–703.

Na Nagara, T. 1972. Effects of no-tillage on soil water use and corn growth on Maury silt loam soil. Master's Thesis, Margaret I. King Library, University of Kentucky.

Na Nagara, T., R. E. Phillips, and J. E. Leggett. 1976. Diffusion and mass flow of nitrate-nitrogen into corn roots grown under field conditions. *Agron. J.* **68**:67–72.

Nelson, L. R., R. N. Gallaher, R. R. Bruce, and M. R. Holmes. 1977. Production of corn and sorghum grain in double cropping systems. *Agron. J.* **69**:41–45.

Peters, D. B. 1960. Relative magnitude of evaporation and transpiration. *Agron. J.* **52**:536–538.

Phillips, R. E. 1974. Soil water and soil temperature. *Proc. No-Tillage Research Conference, University of Kentucky.* Ronald E. Phillips, Editor.

Phillips, R. E., C. R. Belcher, R. L. Blevins, H. G. Miller, G. W. Thomas, T. Na Nagara, V. Quisenberry, and M. McMahon. 1971. *Agronomy Research. Misc. Publication 394.* Kentucky Agric. Exp. Station, Lexington, p. 38.

Quisenberry, V. L. and R. E. Phillips. 1976. Percolation of surface applied water in the field. *Soil Sci. Soc. Am. J.* **40**:484–489.

Quisenberry, V. L. and R. E. Phillips. 1978. Displacement of soil water by simulated rainfall. *Soil Sci. Soc. Am. J.* **42**:675–679.

Radcliffe, D., T. Hayden, K. Watson, P. Crowley, and R. E. Phillips. 1980. Simulation of soil water within the root zone of a corn crop. *Agron. J.* **72**:19–24.

Ritchie, Joe T. 1972. Model for predicting evaporation for a row crop with incomplete cover. *Water Resour.* **8**:1204–1213.

Ritchie, J. T., D. E. Kissel, and Earl Burnett. 1972. Water movement in undisturbed swelling clay soil. *Soil Sci. Soc. Am. Proc.* **36**:874–879.

Russel, J. C. 1939. The effect of surface cover on soil moisture losses by evaporation. *Soil Sci. Soc. Am. Proc.* **4**:65–70.

Shuford, J. W., D. D. Fritton, and D. E. Baker. 1977. Nitrate-nitrogen and chloride movement through undisturbed field soil. *J. Environ. Qual.* **6**:255–259.

Thomas, G. W., R. L. Blevins, R. E. Phillips, and M. A. McMahon. 1973. Effect of killed sod mulch on nitrate movement and corn yield. *Agron. J.* **65**:736–739.

Thomas, G. W. and R. E. Phillips. 1979. Consequences of water movement in macropores. *J. Environ. Qual.* **8**:149–152.

Thomas, G. W., R. E. Phillips, and W. L. Quisenberry. 1978. Characterization of water displacement in soils using simple chromatographic theory. *J. Soil Sci.* **29**:32–37.

Triplett, G. B., Jr., D. M. Van Doren, Jr., and S. W. Bone. 1973. An evaluation of Ohio soils in relation to no-tillage corn production. *Res. Bul. 1068.* Ohio Agric. Res. Dev. Center, Wooster, Ohio.

Triplett, G. B., Jr., D. M. Van Doren, Jr., and B. L. Schmidt. 1968. Effect of corn (*Zea mays* L.) stover mulch on no-tillage corn yield and water infiltration. *Agron. J.* **60**:236–239.

Tyler, D. D. and G. W. Thomas. 1977. Lysimeter measurements of nitrate and chloride losses from soil under conventional and no-tillage corn. *J. Environ. Qual.* **6**:63–66.

Unger, Paul W. 1978. Straw-mulch rate effect on soil water storage and sorghum yield. *Soil Sci. Soc. Am. J.* **42**:486–491.

Wild, A. 1972. Nitrate leaching under bare fallow at a site in northern Nigeria. *J. Soil Sci.* **23**:315–324.

Zartman, R. E. 1974. Transport of nitrate-nitrogen to tobacco roots. Ph.D. Dissertation, Margaret I. King Library, University of Kentucky.

Zartman, R. E., R. E. Phillips, and W. O. Atkinson. 1976. Tillage and nitrogen influence on root densities and yield of burley tobacco. *Tobacco Sci.* **XV**: 136–139.

Zartman, R. E., R. E. Phillips, and J. E. Leggett. 1976. Comparison of simulated and measured nitrogen accumulation in burley tobacco. *Agron. J.* **68**:406–410.

5
Fertilization and Liming

Grant W. Thomas
Professor of Agronomy
University of Kentucky

and

Wilbur W. Frye
Associate Professor of Agronomy
University of Kentucky

THE SOIL ENVIRONMENT UNDER NO-TILLAGE VERSUS CONVENTIONAL TILLAGE

Fertilizing and liming crops grown under no-tillage is not radically changed when compared to conventional practices. There are differences, however. Most of these differences arise from the fact that, under no tillage, the soil is not moved nor disturbed except in the slot where the seeds are placed. Also, under most systems of no-tillage, a residue of dead plant material is left on the soil surface and a kind of natural mulch is formed. The non-disturbance of soil resembles conditions in permanent pasture so that principles proved there are valid for row crops grown under no-tillage in general. The presence of a surface mulch changes the soil water regime, particularly at the soil surface. Principles of fertilization and liming for no-tillage are based on these two conditions as well as on the nutrient requirements of plants and the specific soil and climatic conditions encountered.

When the moldboard plow is used, crop residues, immobile nutrients and most agricultural chemicals are buried to a depth of 6–30 cm, and a new soil "working" surface is created. This new surface is subsequently inverted so that, over a period of years, plant residues, phosphorus fertilizer, lime and herbicide and insecticide residues are pretty thoroughly mixed in the volume of soil between the surface and the general depth of plowing. Any soil which has been farmed conventionally over a period of years exhibits this clearly by its so-called Ap horizon, a dark, uniform surface cap, 15–25 cm in depth.

Soil analyses of organic matter, available phosphorus, calcium, potassium, magnesium and pH show a high degree of uniformity. Soluble nitrate does not show uniformity for reasons that will be discussed later. Uniformity of organic matter and nutrients suggests that the entire zone is an ideal rooting medium which, in turn, suggests that deeper and deeper plowing will enhance yields. These analyses also indicate that a uniform, adequately deep rooting layer is necessary for optimum plant growth. If such is the case, it would be difficult to understand how alfalfa, a very demanding plant, makes excellent growth under conditions where soil properties change drastically with depth.

To contrast with the above situation of a uniform Ap horizon, consider what happens with sustained no-tillage. In this case, all residues return to the soil surface. All fertilizers, lime and chemicals are applied to the soil surface and none are moved unless they dissolve in the soil water. The immediate result is that organic matter and phosphorus are increased markedly in the soil very near the surface and that the deeper portions of the soil are gradually diminished in nutrients. If potassium is applied as fertilizer or if plant residues are high in potassium, it too will accumulate near the soil surface. Calcium and magnesium will decrease near the surface, being solubilized or (more correctly) exchanged by the nitric acid formed from nitrification of either fertilizer or mineralized ammonium. If a regular liming program is followed, however, calcium and magnesium will be very high near the surface and decrease exponentially with depth. Soil analyses showing these features are presented in Table 5-1.

In this response to the surface concentration of nutrients and water, one can expect that plant roots will proliferate in that shallow zone. This is, indeed, the case (Table 5-2). It is also true that soil microorganisms are very highly concentrated in the same shallow zone (Table 5-3). No doubt worrying about impending drought, many thoughtful people have suggested that occasional tillage be practiced to place some of the nutrients, organic matter and micro-

Table 5-1. Soil properties under limed and unlimed and tillage treatments. (From Blevins *et al.*, 1982.)

	NO-TILLAGE									
	CALCIUM (meq/100 g)		MAGNESIUM (meq/100 g)		POTASSIUM (meq/100 g)		ORGANIC CARBON (%)		PHOSPHORUS[1] (ppm)	
DEPTH (cm)	LIMED	UNLIMED	LIMED	UNLIMED	LIMED	UNLIMED	NO-TILL	CONVEN-TIONAL	NO-TILL	CONVEN-TIONAL
0–5	9.77	2.61	0.76	0.47	0.59	0.44	2.79	1.37	242	63
5–15	7.50	4.74	0.58	0.55	0.32	0.34	1.31	1.31	21 12	44 37
15–30	7.15	6.10	0.59	0.54	0.20	0.20	0.70	0.75	8	17

[1] Shear and Moschler (1969), 129 ppm applied in six years.

Table 5-2. Corn root density with soil depth under conventional and no-tillage. (From Phillips *et al.,* 1971.)

DEPTH (cm)	ROOT DENSITY (cm of root/cm^3 of soil)	
	CONVENTIONAL	NO-TILLAGE
0–5	1.8	21.1
5–15	6.0	5.2
15–30	5.5	1.8

Table 5-3. Microbial ratios between no-tillage and conventional soils. (Reproduced with the permission of John W. Doran, Soil Scientist, Agricultural Research Service, U.S. Department of Agriculture, Lincoln, Nebraska, from *Soil Science Society of America Journal* **44:765–771 [1980].)**

NO-TILLAGE/CONVENTIONAL					
DEPTH cm	TOTAL AEROBES	NH4 OXIDIZER	FUNGI	FACULTATIVE ANAEROBES	DENITRIFIERS
0–7.5	1.35	1.25	1.57	1.57	7.31
7.5–15	0.71	0.55	0.76	1.23	1.77
0–15	1.03	0.89	1.18	1.32	2.83

organisms deeper, encouraging deeper root penetration. However, there is no real evidence that there is a crop advantage to deeper placement of nutrients in a no-tillage system.

The surface of a mulched soil is the wettest part of the soil during much of the growing season. Roots extract water first from that part of the soil where water is easiest to extract (the wettest part) and having the nutrients concentrated in that same volume of soil appears to be an ideal situation for optimum nutrient uptake, especially during early stages of crop growth. If, later in the season, the water is exhausted from the soil surface, the roots will remove water from depths where soil water is more available. Nutrient uptake may suffer at times but the heaviest rate of nutrient absorption has usually passed. Specific examples for various nutrients will follow.

Other differences in soil environments between soil under no-tillage and conventional tillage, both related to the surface mulch, are the tendency for the no-tillage soil to be both wetter and colder. The wetter surface is thought to be due both to reduced evaporation and often to an improved infiltration of rainfall. The cooler soil temperatures are related to the insulating effect of the mulch, the greater path of diffusion of water vapor through the mulch and the (usual) change in surface albedo which tends to reflect radiation rather than to absorb it. These points are discussed at length in Chapters 2 and 4.

Another major difference between no-tillage and conventional soils is that crop residues are not mixed thoroughly with the soil. This delays their decomposition (Chapter 9) and affects nitrogen availability to crops, usually negatively.

In summary, the upper soil under no-tillage generally is wetter, cooler, higher in organic matter, microorganisms, phosphorus and potassium and somewhat higher in acid than a comparable soil that is tilled.

Nitrogen Requirements for No-Tillage Crops

The actual nitrogen uptake by plants grown under conventional and no-tillage does not vary except with crop yield. If yields are identical, there will be no difference in nitrogen requirements. If, as often happens, no-tillage crops yield 10 percent more, the nitrogen requirement will be 10 percent higher. Hence, the crop itself does not require any different amount of nitrogen to perform well in the two systems. The difference in nitrogen fertilization requirements arises either from expected yield differences or from soil reactions which are different under conventional compared to no-tillage conditions.

Pathways of Nitrogen in the Two Systems

Behavior of nitrogen under no-tillage is modified somewhat compared to its behavior under conventional tillage due to differences in the soil environment. The nitrogen mineralization rate tends to be lower because the soil is not disturbed and the organic residues remain on the surface where decomposition is slower. Hence, especially in the spring, there is usually less nitrate in the soil in unfertilized no-tillage soil than in similar tilled soil (Rice and Smith, 1982; Thomas *et al.*, 1973).

In the case of biological denitrification, Rice and Smith (1982) have found that the denitrification rate in undisturbed soil cores is linearly related to soil water content. Because the water content of untilled and mulched soils, especially in the surface, is higher, the rate of biological denitrification can be expected to be higher than in tilled soil. Depending on the "wetness" or drainage class of the soil, this effect of no-tillage can be small (well-drained soils) or very large (poorly-drained soils). In general, it is a noticeable problem whenever a soil is less than well-drained.

Leaching of nitrate is also favored under no-tillage, particularly during the spring and early summer, before a crop canopy is formed. In general, leaching of nitrate occurs when evapotranspiration is exceeded by rainfall. In Kentucky and in much of the humid region of the United States, this period extends from November through May. Beginning in May or June, the amount of water lost via evapotranspiration is greater than the amount of rainfall. The effect is shown for Lynchburg, Virginia in Figure 5-1. However, under a mulched no-tillage system with small plants, the loss of water by evapotranspiration is considerably less (Hill and Blevins, 1973) than that from a conventionally tilled soil (Table 5-4). The lower evaporation from no-tillage soil during that period causes the potential leaching of nitrate to be extended well into June rather than May. Similar differences will be noted in other climatic regions due to the

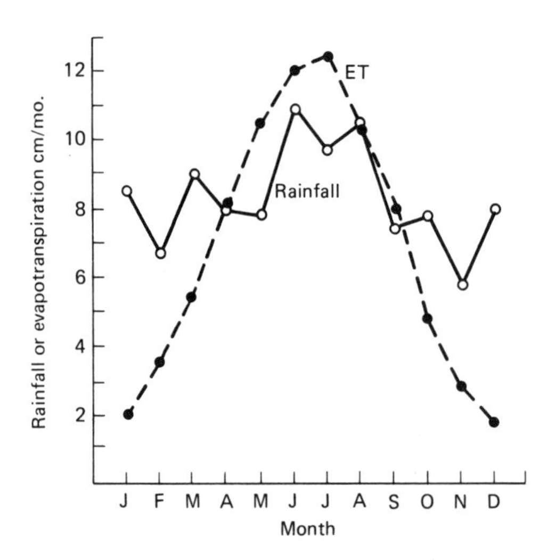

Figure 5-1. Rainfall and evapotranspiration for Lynchburg, Virginia (from Van Bavel and Lillard, 1957).

Table 5-4. Effect of tillage on early season water loss by evapotranspiration (ET) at Lexington, Kentucky. (From Hill and Blevins, 1973.)

	ET FROM PLANTING TO JUNE 20 (cm)		
TILLAGE	1969	1970	1971
No-tillage	11.6	13.8	11.8
Conventional	14.7	14.9	13.7
Difference	3.1	1.1	1.9

surface mulch. These differences in evaporation from mulched and bare soils are discussed in greater detail in Chapter 4.

Another property of soils under no-tillage also tends to promote nitrate leaching. When an undisturbed (no-tilled) soil receives rain, the water tends to move down the larger pores rather than through the bulk of the soil. Larger pores which are not broken by soil tillage can extend to great depth in well-structured soils. Therefore, high-intensity rains cause losses of nitrate from untilled soils when tilled soils do not lose any at all (Thomas *et al.*, 1973; Tyler and Thomas, 1977).

Taken together, every reaction that can cause lower available nitrogen is favored by no-tillage. It is impossible to say, in a given soil, which of the reactions (slower mineralization, higher denitrification or greater leaching) is most important. However, denitrification appears to be a greater problem in imperfectly-drained soils and leaching tends to be more important in well-drained soils. As a rule, mineralization is slower in any soil that is cooler, wetter and undisturbed, as in no-tillage.

Rather early in our research, we were faced with poorer nitrogen efficiency under no-tillage. In addition to determining the reasons for this we also had a strong interest in adopting a practical means of overcoming the disadvantage. Table 5-5 shows the result of three years work on a Hampshire soil near Frankfort, Kentucky, a soil with rather slow drainage because of a clay subsoil. These results show clearly that delaying nitrogen fertilizer until a month after planting corn gives much higher efficiency than fertilization at planting time. Since this observation, we have had many field experiments on time of nitrogen application to no-tillage corn on a very wide range of soils, using both

Table 5-5. Effect of time of nitrogen application on corn yield on a Hampshire silt loam. (From Miller *et al.*, 1975)

N RATE (kg/ha)	YIELD (kg/ha) 1972	1973	1974	AVERAGE
0	5,700	4,900	3,700	4,767
168 at planting	6,300	6,650	6,550	6,500
168–delayed 4 weeks	7,700	8,900	7,950	8,063

conventional and no-tillage corn. The results are summarized in Figure 5-2.

In general, conventional corn yields were not increased by delaying or splitting nitrogen application, whereas no-tillage corn yields increased most with a split application and considerably with a delayed application. At the time the research reported in Table 5-5 was begun, nitrogen fertilizer was inexpensive and farmer acceptance of delayed or split application of nitrogen fertilizer was poor. The rapidly increasing price of nitrogen has changed this markedly; however, split application of nitrogen is now accepted with no-tillage corn in much of Virginia, Maryland, Delaware, Indiana, Ohio, Kentucky and southern Illinois.

If it is true that delayed application of nitrogen helps avoid early losses due to denitrification and leaching, it also is true that mineralization will not be changed and that no-tillage corn without fertilizer should give lower yields. This has proven to be true in most cases

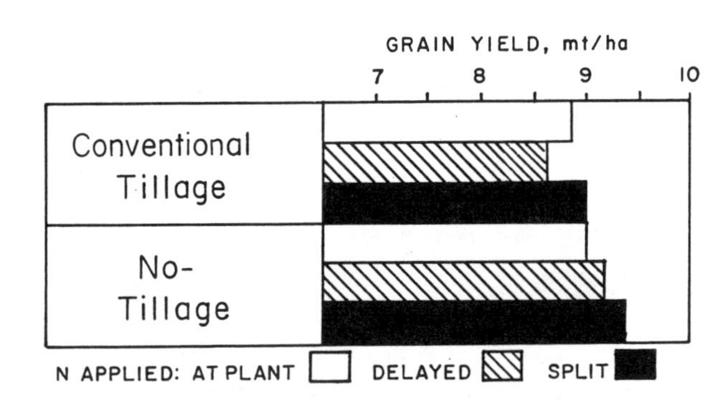

Figure 5-2. Effect of time of N application on corn yield and tillage (from Frye *et al.*, 1979).

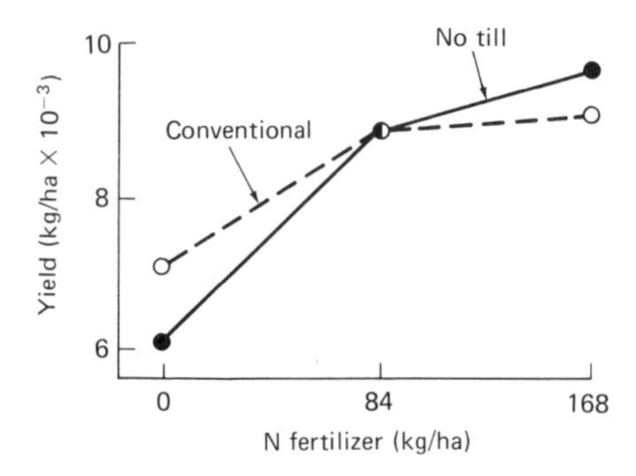

Figure 5-3. Response of corn to nitrogen fertilization; values are the average of eight locations in Kentucky (reprinted from June 1970 *Science* 208:1108-1113, Phillips *et al.*, copyright 1980 by the American Association of the Advancement of Science).

under the climate and soils of the humid region middle latitudes (Phillips *et al.*, 1980) (Figure 5-3), but this depends strongly on rainfall patterns and soil characteristics. For example, in each of the preceding eleven years, we have observed lower yields on check plots of no-tillage corn compared to conventional corn. However in 1981, a year having a wet spring after a drought, the conventional corn yielded the same as the no-tillage corn indicating that some of the nitrogen mineralized early was lost from the conventionally-tilled soil and that the no-tillage corn caught up with it due to slow but steady mineralization.

An even more extreme case of this behavior was observed in Costa Rica by Shenk and Saunders (1980). The soils they were working with are high in organic matter (12 percent) and rain occurs nearly every day. Their results show much better yields with no-tillage corn (unfertilized) than with conventional corn (Table 5-6). In different

Table 5-6. Effect of nitrogen rate on corn yield under no-tillage and conventional tillage in Costa Rica. (From Shenk and Saunders, 1980.)

N RATE (kg/ha)	YIELD (kg/ha)	
	NO-TILLAGE	CONVENTIONAL
0	2,420	1,800
40	2,570	2,500
80	2,610	2,450
120	2,690	2,830

climatic regions and with different soils, many different results can be expected, but the principles will remain unchanged.

Methods of Nitrogen Application

Almost every method of nitrogen application has been tried with no-tillage crops. The most common practices are surface broadcasting of solid nitrogen fertilizer, spraying of liquid nitrogen fertilizer and knifing in anhydrous ammonia. All these are used by successful no-tillage farmers. The broadcasting of solid fertilizer nitrogen is the cheapest method of application but uses the most expensive materials. In general, it is suitable for moderate acreages where anhydrous ammonia use is impractical. When ammonium nitrate is used, there is no volatilization, but denitrification and/or leaching are practical problems, depending on soil water content and rains soon after application.

Urea use generally means some degree of loss by the reaction with urease and subsequent volatilization of ammonia. Losses are in the range of 0–20 percent. Surface application of liquid nitrogen, usually urea-ammonium nitrate solutions, is commonly combined with the herbicide spraying in order to increase the killing effect of the herbicide and to save a trip over the field. The urea portion of the solution is again subject to ammonia volatilization, perhaps even more than with solid urea, since contact with plant residues (and urease) is more likely. Liquid ammonium nitrate does not appear to be affected any differently from the solid form. Overall volatilization losses are in the range of 0–15 percent but probably average 10 percent or less.

Anhydrous ammonia application is perfectly feasible with no-tillage but requires some care because the surface soil is usually wetter than with conventional tillage. Under this condition and without dry particles to seal the slits, anhydrous ammonia can be lost by volatilization. As with conventional tillage, care is required to assure sealing of the slits after anhydrous ammonia is applied with no-tillage.

If any of these three methods of nitrogen application is used for sidedressing, its relative advantage changes somewhat. For example, surface broadcast application of solid fertilizer will cause plant leaf burning and some plants will be run over. In our experience, there is not an economic effect on corn yield due to burning or smashing, but there is a temporary aesthetic setback. Spraying liquid nitrogen usually has to be modified to "dribbling" so that plants are not badly burned. With anhydrous ammonia application as a sidedress, there is less chance of loss because the soil is likely to be drier and easier to seal.

The question of nitrogen placement in no-tillage has been raised recently in Maryland, where Bandel (1981) has obtained excellent response from banding a part of the required nitrogen early and the rest later.

Similarly, in Alabama, Touchton (1982) has increased yields by use of a complete "pop up" fertilizer. In this latter case, the effect probably is an effect of both nitrogen and phosphorus, or, perhaps, is mostly due to phosphorus.

In no-tillage crops, nitrogen from the soil itself is somewhat less available to plants in general. This appears to be due to less mineralization, more denitrification and more leaching. Added nitrogen suffers the risk of more denitrification and more leaching. However, at the same time, the potential for maximum yield is often higher with no-tillage and therefore nitrogen uptake by the crop is often higher (see Figure 5-3). Making sure that the nitrogen supply to crops is adequate under no-tillage usually means adding a greater total amount or using means such as sidedressing or placement to minimize loss. Obviously, the latter approach is more energy-efficient and most often costs less as well.

Phosphorus Requirements

The requirements for phosphorus by crops grown under no-tillage will be as great as or slightly greater than for conventionally grown crops. If the requirement is greater, it will be only because of slightly higher yields with no-tillage. In general there will be no significant differences in requirements due to uptake alone.

In soils, there are three factors affected by no-tillage which in turn strongly affect uptake of phosphorus by plants. The first of these is soil temperature. In general, under a mulched surface, the mean temperature is lower than it is where a soil has been disturbed such as in conventional tillage. When this mean temperature is below the critical temperature for seedling growth and development (about 10°C for corn), phosphorus uptake will be inhibited. Thus, if no-tillage corn is growing in a soil with a mean temperature of 14°C and conventionally planted corn is found in a soil with a mean temperature of 17°C, better phosphorus uptake can be expected by the conventional corn, all other things being equal. As in the case of conventional corn, addition of phosphorus in a band can be expected to overcome at least some of this temperature effect and promote early growth. Touchton (1982) in Alabama has shown a marked response to banded fertilizer with no-tillage corn when it is planted earlier than normal to escape summer drought.

The second factor influenced by tillage and influencing phosphorus uptake is the soil water content. The amount of phosphorus in most soil solutions is so small that uptake by mass flow is almost insignificant. Therefore, most uptake depends on the diffusion of phosphorus. Diffusion of phosphorus, in turn, is influenced strongly by the soil water content because phosphorus ions diffusing from one point to another require continuous films of water (Olsen and Watanabe, 1963). The effect of water content on ^{32}P diffusion is shown in Figure 5-4 (Mahtab *et al.*, 1971).

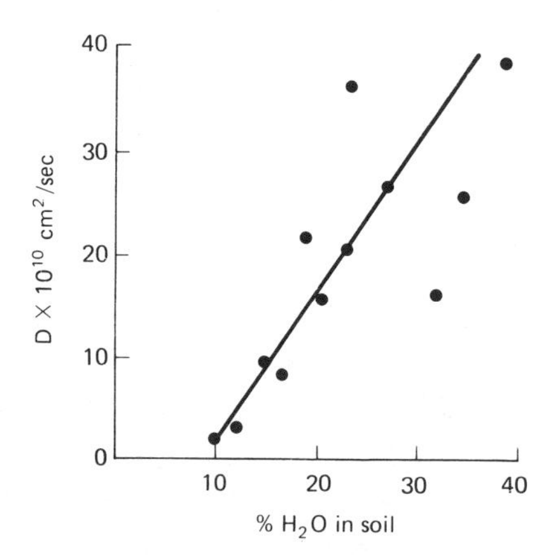

Figure 5-4. Effect of soil water content on P difusion coefficient of four Texas soils (from Mahtab *et al.*, 1971).

As stated earlier, one of the chief distinguishing features of a soil under no-tillage is the higher content of soil water, especially near the soil surface. Therefore it would be expected that under no-tillage, phosphorus diffusion rate would be higher than in the conventional soil. Early studies made by Singh *et al.* (1966) suggest that this is the case. Table 5-7 shows data for ^{32}P and total phosphorus uptake by corn from that paper. Notice that ^{32}P uptake was considerably higher under no-tillage than under conventional tillage. Total phosphorus uptake also was higher early in this season. Even more significantly, early growth was much greater even on the no-tillage check than on the conventional check and final yield was higher than the fertilized conventional corn and twice as high as the unfertilized conventional corn. Temperature differences were rather small and since both treatments were above a mean temperature of 15°C, temperature probably had little effect on phosphorus uptake.

The third factor influencing phosphorus uptake that is affected by tillage is the placement of phosphorus fertilizer. In the experiment cited above, the labeled phosphorus was mixed uniformly to a depth of 12.5 cm in the conventional plots and spread on the soil surface in the case of the no-tillage plots. In a sense, surface application can be seen as a sort of "horizontal banding" of phosphorus because a minimum of soil contact occurs. Hence, a longer period of availability would be expected than when fertilizer phosphorus is mixed with the soil. In an attempt to check this point, Belcher and Ragland (1972) conducted an experiment on a very low-phosphorus soil where phosphorus was either placed entirely on the soil surface or was

Table 5-7. Percent P from fertilizer and percent P in corn plants. (From Singh *et al.*, 1966.)

DAYS AFTER PLANTING	% P FROM FERTILIZER		% P IN PLANT	
	NO-TILLAGE	CONVENTIONAL	NO-TILLAGE	CONVENTIONAL
30	54	16	0.07	0.04
46	43	32	0.18	0.18
60	25	21	0.16	0.13
67	36	37	0.15	0.15

Table 5-8. Effect of surface vs. surface plus banded P application on corn yield. (From Belcher and Ragland, 1972.)

RATE OF P (kg/ha)	CORN YIELD (kg/ha)	
	SURFACE	SURFACE + BANDED
0	4,645	4,771
56	7,533	5,901
112	6,278	6,152
224	6,528	6,780

partly banded and partly surface-applied. Only no-tillage was used. The results (Table 5-8) show that yields were as good as or better than partially banded treatments when all phosphorus fertilizer was applied to the surface. Results showing superior uptake of P (and K) placed on the soil surface also were obtained by Triplett and VanDoren (1969) in Ohio (Table 5-9).

Naturally, the surface application of phosphorus year after year increases the phosphorus concentration in the soil surface but does little for the deeper layers. Hence, it is at least plausible that the method of soil sampling should be altered as it is in the case of pastures, to reflect the fertilizer effect more precisely. In Kentucky, farmers are advised to sample to a 10-cm depth for no-tillage versus a 15-cm depth for conventionally tilled crops.

While it is true that phosphorus research under no-tillage is of very short duration and therefore has not had nearly the evidence that con-

Table 5-9. Effect of tillage on P and K in corn plants. (From Triplett and VanDoren, 1969.)

NUTRIENT	8–10 LEAF STAGE		AFTER TASSELING	
	CONVENTIONAL	NO-TILLAGE	CONVENTIONAL	NO-TILLAGE
	(% in plant tissue)			
P	0.29	0.35	0.27	0.27
K	3.92	4.20	2.00	2.06

ventional tillage has behind its recommendations, it also is true that the results have been more consistent and that farmers have been successful in following recommendations based on these results. It would be an oversimplification to say that surface application of phosphorus gives the best results in every instance but it does work in a vast majority of soils and climatic conditions. An outline of possible exceptions follows.

1. *Soils with mean temperature <15°C at planting.* Starter phosphorus may give positive results; final yield may or may not be affected.
2. *Soils with high soil test P level in surface 5 cm and low soil test P below that level.* Opinion is divided. Most studies indicated that phosphorus uptake from deeper soil levels is not required. However, persistent drought *could* reduce P uptake from the soil surface and for insurance an occasional deep P placement or plowing could be used.
3. *Soils with very low soil test phosphorus.* There is no evidence that banding P fertilizer will enhance P uptake above that found by surface application under no-tillage.

Phosphorus rates for no-tillage have not been studied in any detail. Where comparisons have been made there is either no difference in response or the response curve flattens out earlier with no-tillage. In low-phosphorus soils, the curve shown in Figure 5-5 is rather representative (Moschler, 1971). Notice that the yield is higher with no-tillage when no phosphorus is applied and that the major response is between 0 and 19 kg/ha of applied P. With conventional tillage, however, the response is linear up to the 38 kg/ha rate.

In general, it is possible to conclude that phosphorus uptake and efficiency of phosphorus fertilizer utilization are superior to what is found under conventional tillage. This is a rather encouraging result, because it suggests that on many marginal hillside soils where only no-tillage would be adapted, there will not be any unusual problem with phosphorus fertilization. This, in turn, suggests that the economics of using such soils will be reasonably favorable.

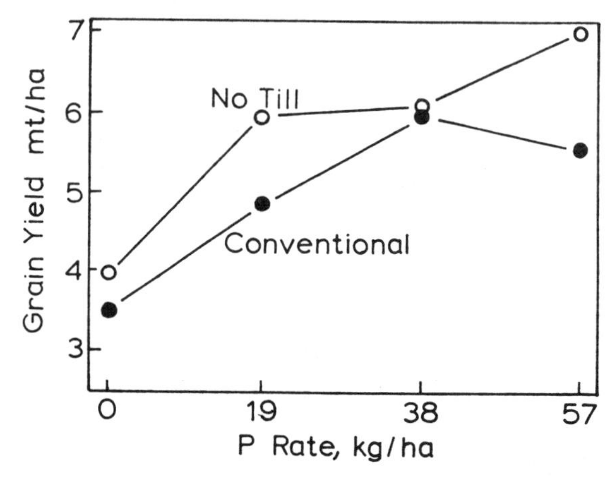

Figure 5-5. Phosphorus effects on corn yield under conventional and no-tillage corn. (Moschler, W.W., 1971)

Potassium Distribution and Availability in the Soil

The behavior of potassium in soils where no-tillage is practiced is not apt to be nearly so different from that observed under conventional tillage as has been noted for nitrogen and phosphorus. Potassium is a monovalent cation which is held rather firmly by the soil, so that when sufficient cation-exchange capacity is present, very little loss of leaching occurs. The one factor which affects potassium distribution under no-tillage more than any other is the lack of soil movement under no-tillage.

When potassium is applied to soils under conventional tillage, it usually is plowed or disked into the soil immediately after fertilization. Then, in subsequent years, it is mixed further as more tillage is done. During this time, too, additional potassium is usually being added. The result is that the zone of tillage is relatively uniform in potassium with a noticeable diminution of available potassium occurring below that depth.

When no-tillage is practiced, however, fertilizer potassium is applied to the surface of the soil and the only movement that occurs is that due to leaching. The distribution of potassium with depth is governed by cation exchange between potassium and the cations already present on the soil (mostly calcium) so that the highest potassium content is encountered at the soil surface followed by a sharp decline in potassium content with depth. Examples of potassium distribution after 10 years of conventional and no-tillage corn are shown in Figure 5-6.

This difference in potassium distribution has not been shown to be important in plant nutrition. It is important in sampling soils for fertility evaluation because the normal 15-cm sampling depth is based on determining the potassium content of the uniform plow layer. Because of this, in Kentucky, it is recommended that soil samples be taken only from a 10-cm depth as already discussed for phosphorus.

There have been scattered reports of potassium deficiencies under no-tillage in soils where "adequate" supplies of available potassium are present according to soil test. This has been attributed to a combination of higher bulk density (at least early in the season) under no-tillage and higher water content, thus reducing the quantity of air-filled pores (Moncrief and Schulte, 1982). The contention could well be true, since most reports occur in cold, wet spring seasons. The effect, however, can be overcome either by having a

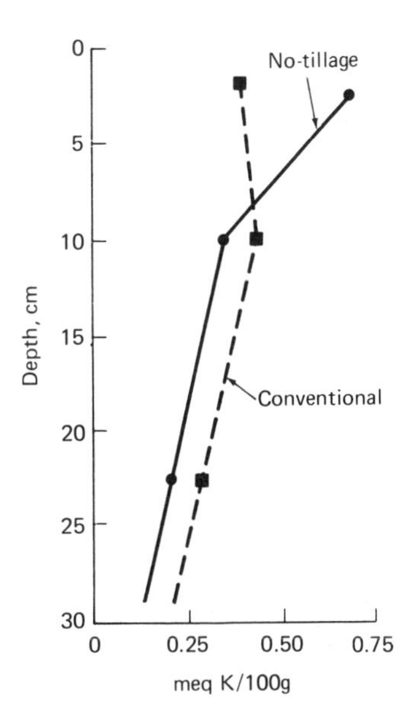

Figure 5-6. Exchangeable K distribution with depth after 10 years of no-tillage and conventionally-planted corn. (Blevins *et al.*, 1983)

high level of potassium in the soil or by addition of small amounts of potassium in the row. Data in Table 5-10 from Moncrief and Schulte (1982) indicate that K uptake is measurably lower under no-tillage but they also found that P uptake was greatly increased.

In Kentucky we observed this phenomenon of K deficiency in one year in a field that had been planted to corn and tobacco in previous years. In the year after which no-tillage corn was planted over the whole area, the tobacco field corn was normal in size and color, while corn following corn was small, spindly and yellow. Analysis of the corn leaves showed that the normal corn was high in potassium and the spindly and yellow corn was low. Soil analyses showed approximately 600 kg/ha of available K in the tobacco field area (normal corn) and 200 kg/ha in the area where the poor corn was growing. However, within two weeks, the differences had largely disappeared and all corn was normal.

It seems apparent that under wet, cold conditions using no-tillage, there is a possibility of induced potassium deficiency on soils where potassium is in the medium-to-high range. This deficiency may or may not persist, depending on weather conditions, and it may or may not have a lasting influence on yield. In the northern part of the U.S., or with very early planting in the southern U.S., the effect is magnified. It is clear that the problem may be overcome by either having a high level of potassium in the soil or by small additions of row fertilizer potassium.

In Kentucky, where most no-tillage corn is grown on well-drained soils and where no particular effort is made to put corn in early, the potassium deficiency problem is almost unknown. With soybeans,

Table 5-10. Uptake of P and K by corn as affected by tillage in a Fayette silt loam, Lancaster, Wisconsin. (From Moncrief and Schulte, 1982.)

	UPTAKE[1] (mg/plant)	
TILLAGE	PHOSPHORUS	POTASSIUM
No-tillage	1.80	12.3
Chisel	1.35	13.8
Moldboard plow	1.22	15.9

[1] At soil test level of "high."

which are planted later and after wheat, few problems should be encountered, and none, so far as the authors are aware, have been reported.

It is probable that, under cold, wet conditions, the possibility of using row potassium fertilizer should be provided for. As an "insurance" precaution, this could be done routinely at a small cost except on fields where soil test potassium is unusually high. With this exception, it appears that potassium is not a problem under no-tillage, but soils and areas where potassium starter and/or high soil potassium are required need to be delineated so that farmers will not experience severe yield losses.

Soil Acidity and Lime

The Soil Acidity Problem. As stated in the introduction, most of the acidity in modern agriculture arises from the nitrification of ammonium nitrogen to nitrite by nitrosomonas bacteria.

$$2NH_4^+ + 3O_2 \longrightarrow 2NO_2^- + 2H_2O + 4H^+$$

The second step of the reaction, however, produces no further acidity.

$$2NO_2^- + O_2 \longrightarrow 2NO_3^-$$

Because for each NH_4^+ oxidized to NO_2^-, there are two H^+ formed, the total *potential* acidity from each 100 kg/ha of ammonium nitrogen added is equivalent to $\frac{100\ \text{kg}}{14\ \text{g N/mole}} \times 2$ equiv H/mole $\times$ 50 g/equiv $CaCO_3$ = 700 kg $CaCO_3$/ha. In practice, it is seldom this high because of plant uptake of ammonium, nitrogen losses from denitrification, etc. But it clearly is more important than acidification from natural rainwater or even acid rain. With H_2CO_3 from "clean" rainwater present at a concentration of 0.012 meq/liter, the potential acidification on a hectare is only 6 kg of $CaCO_3$ equivalent with a 1,000-mm yearly rainfall. The rainfall would have to be 100 times as acid in natural rainfall to approach the acidification from nitrogen, whereas it probably is typically no more than 10 times as acid.

Table 5-11. pH of 0-5 and 5-15 cm soil layers after 10 years of nitrogen application as affected by tillage. (From Blevins *et al.*, 1982.)

	pH AFTER 10 YEARS			
	NO-TILLAGE		CONVENTIONAL TILLAGE	
YEARLY N RATE (kg/ha)	0-5 cm	5-15 cm	0-5 cm	5-15 cm
0	5.75	6.05	6.45	6.45
84	5.20	5.90	6.40	6.35
168	4.82	5.63	5.85	5.83
336	4.45	4.88	5.58	5.43

With no-tillage, the amount of leaching of nitrate and the basic cations accompanying it is apt to be slightly higher than with conventional tillage. This difference, however, has only marginal effects on acidity. The major difference is caused by application of nitrogen fertilizers to the soil surface which is not mixed with other portions of the soil. The long-term effects of tillage and nitrogen on soil pH are shown in Table 5-11. It is clear from these data that surface acidification of soil under no-tillage can be quite severe, especially when high nitrogen rates are used. This, in turn, has serious consequences on release of toxic elements and efficiency of herbicides of the triazine group.

When soil pH drops, two toxic elements, aluminum and manganese, become problems for plant growth in most soils. In the Maury silt loam in which we have compared conventional and no-tillage corn for the past 12 years, both of these elements showed a striking increase with time, especially under no-tillage. The data for 5 and 10 years are given in Table 5-12 for aluminum and manganese. Note that although surface pH did not vary much between 5 and 10 years, the exchangeable aluminum level essentially doubled. The reason for this slow and gradual increase in aluminum appears to be the high content of organic matter in the surface of the Maury soil. Thomas (1975) showed that the amount of exchangeable aluminum in the Maury soil, at a given pH, was negatively related to percent organic matter. Further work by Hargrove and Thomas (1981) demonstrated that organic matter "protected" acid soils to some extent from high levels of aluminum by complexing it strongly.

Table 5-12. Al and Mn in the 0-5-cm layer after 5 and 10 years of no-tillage at 168 and 336 kg/ha N applied each year. (From Blevins *et al.*, 1977 and 1982.)

	5 YEARS	10 YEARS
Al, meq/100 g		
168 kg/ha N	0.32	1.22
336 kg/ha N	1.08	2.00
Mn, μg/g		
168 kg/ha N	7.3	28.0
336 kg/ha N	19.9	52.5

This indicates that toxic aluminum concentrations in soils under no-tillage are lower where organic matter levels are high. Since organic matter levels climb under no-tillage, there is some compensation for one of the effects of soil acidity. In the case of manganese, however, the organic matter does not complex the cation strongly. Hence, there is no protective effect of organic matter on manganese toxicity (Hargrove and Thomas, 1981).

The triazine herbicides appear to be easily hydroxylated in acid soils, rendering them ineffective as weed killers (Best *et al.*, 1975). Work done on the Maury soil at Lexington, Kentucky (Kells *et al.*, 1980) showed this effect strongly. On the most acid plots weed control was extremely poor (Table 5-13) and, in fact, had a larger effect on reducing yields than did either aluminum or manganese toxicity. Grass competed for light, water and nutrients and much of the corn failed to mature normally. Of all the negative effects of soil

Table 5-13. The effect of pH on percent weed control in corn. (From Kells *et al.*, 1980.)

pH	% WEED CONTROL
5.5	62
5.2	54
4.7	48
4.3	19

acidity, it appears that poor weed control may show up the soonest and possibly give the most severe loss of yield.

Correcting Soil Acidity. The least expensive and most straightforward way to correct soil acidity is by applying ground limestone to the soil. As a rule, limestone is applied at rates of from four to eight tons per hectare and plowed and/or disked into the soil. This tillage and subsequent tillage provides fairly complete mixing of lime with the soil so that maximum effect is obtained quickly. When no-tillage is practiced, the opportunity to incorporate lime does not exist and the lime reaction is not as efficient as when lime is mixed thoroughly with the soil.

In a study conducted at two sites in Kentucky (Blevins *et al.*, 1978), on Tilsit and Maury silt loams, lime was applied only to the soil surface in the no-tillage plots and never mixed. In the conventional plots, the lime was mixed with the soil by plowing and disking. Results of liming on pH for the Tilsit sil are shown in Table 5-14. The surface application of lime at a rate of 10.1 mt/ha increased the pH at every level down to a depth of 30 cm with no-tillage, suggesting that measurable quantities of lime and/or $Ca(HCO_3)_2$ reacted to a depth of 30 cm. Where the lime was mixed to a 15- or 20-cm depth (10.1 mt/ha), the increase in pH at the 10–30-cm depth was 0.5 of a

Table 5-14. Soil pH of Tilsit silt loam soil after five years from different lime rates on no-tillage and conventionally tilled corn. (Adapted with the permission of Blevins *et al.* from *Agronomy Journal* 70, *2*:323 [March–April 1978].)

	LIME RATE (mt/ha)					
	NO-TILLAGE			CONVENTIONAL		
SOIL DEPTH	0	3.4	10.1	0	3.4	10.1
(cm)			(pH)			
0–5	4.6	5.5	6.4	5.2	5.5	5.7
5–10	5.5	5.7	6.4	5.4	6.0	6.5
10–20	5.5	5.6	5.9	5.6	5.8	6.6
20–30	5.0	5.1	5.2	5.2	5.3	5.5

unit more than it was in the case of the surface applied lime; however, in the surface 5 cm, the pH was 0.7 of a unit more for the surface applied lime (no-tillage) than it was where the lime was mixed to a depth of 15 or 20 cm (conventional). From this experimental work and from years of experience in liming permanent pastures, we have concluded that surface liming of no-tillage soil can be accomplished quite satisfactorily.

A recent study from Georgia (Hargrove *et al.*, 1982) and one from Kentucky (Kitur, 1982) suggest that care must be taken with this recommendation. Kitur found that lime coarser than 60 mesh applied to the soil surface did not react measurably over a period of 40 days. The <60 mesh limestone reacted roughly half as fast when placed on the surface as did the mixed limestone. The results of Kitur suggest that if one is dealing with a very acid soil which needs rapid pH change, it may be more efficient to mix the lime thoroughly with the soil. Thereafter, surface application of lime should do a satisfactory job. The Georgia work of Hargrove *et al.* (1982) shows a small increase in acidity with time at lower depths and the possibility of some need for mixing lime. In Kentucky, Blevins *et al.* (1978) achieved the desired result from either a single large surface application or a series of small, yearly surface applications.

Except in cases of extremely acid soils, surface application of lime to soils which are not tilled appears to perform satisfactorily. In addition, a crop rotation where soybeans are used greatly reduces the need for lime because they are not fertilized with nitrogen. Wheat is fertilized with nitrogen but at a rate of about half that used in corn. It does appear, however, that when surface application of lime is used, care should be taken to obtain a limestone that is ground rather finely. Coarse limestone is ineffective when placed on the surface.

Sulfur

The sources of sulfur for plants are essentially four: irrigation water, fertilizers, rainwater and atmospheric fallout and soil organic matter. Most irrigation water is high enough in sulfur that it will provide more than enough for plant needs. Fertilizers which formerly contained high levels of sulfur are not nearly so well-supplied as at the present now that triple superphosphate and ammonium phosphates have largely supplanted ordinary superphosphate, at least in the United States. The principal fertilizer which contains significant quantities of sulfur is ammonium sulfate, which is used in only limited parts of the world. If it is used, however, there should be no need for any further sulfur. Sulfur supplied through rainfall and atmospheric pollution is sufficient, in many parts of the U.S., to supply all the plant needs for this element. In other parts of the country, where industrial pollution is not significant, the content of sulfur in air and rainfall is far below that required for plant growth. Not only are there large regions of the country where atmospheric sulfur is low (the Northwest and the Deep South), but there are small areas within states where sulfur pollution is unusually low.

A national study on acid rain will soon be generating up-to-date information on sulfur levels in all of the U.S. A classic earlier study (Jordan *et al.*, 1959) for the Southeast was conducted at a time when coal burning patterns were vastly different than what they are today. A smattering of recent information, such as that by Jones *et al.* (1979) indicates rising sulfur levels in the atmosphere for formerly "clean" areas.

In areas where atmospheric sulfur is low and where irrigation and fertilization are not contributing to sulfur supply, the organic matter content of the soil is the chief source of sulfur. Where this is the case, the effect of tillage on the sulfur available to plants can be large, indeed.

When soil is tilled, the breakdown rate of organic matter is increased tremendously for a short time. As organic matter is decomposed, nitrogen, phosphorus and sulfur are mineralized at a rate which roughly corresponds to the relative quantities present. It is not unusual to have N:S ratios of 10 or 20 to 1 in organic matter so that sulfur mineralized is adequate for the amount of nitrogen *mineralized* (the usual N:S ratio in plants is about 15:1) but not necessarily adequate for *total* nitrogen uptake. With no-tillage this quantity is reduced even further, so that chances of sulfur deficiency are greatly enhanced on soils which are marginal in sulfur when ordinary tillage is practiced.

There have been reports of sulfur deficiency under no-tillage when no deficiency was observed with conventional tillage (Koehler *et al.*, 1977). The extent of sulfur deficiency due to no-tillage is not known but in any area where sulfur deficiency is suspected it would very likely be worse under no-tillage just as nitrogen deficiency (with no nitrogen applied) is usually worse with no-tillage.

Treating the sulfur deficiency does not seem to be prohibitive in cost. Gypsum is an excellent source of sulfur and is practically free for the asking in many parts of the country. Where gypsum is not available, fertilizers formulated with sulfur can be used. The total sulfur that needs to be added seldom exceeds 30 kg/ha except on soils that hold sulfate tightly such as andosols.

Micronutrients

There is so little evidence on availability of micronutrients as affected by tillage that only a few speculations will be made here. The first relationship that comes to mind is the generally higher organic matter and reduced erosion under no-tillage. This should favor the availability of zinc. The higher water content of the upper part of the soil profile could either enhance micronutrient uptake when it favors root growth or hamper micronutrient uptake when it is too wet and reduces root growth. Over most of the season, however, micronutrient uptake should be enhanced. Acidification of the surface should improve micronutrient availability except in the case of molybdenum, while surface liming should result in less availability of zinc and manganese.

In Kentucky, we have never noted *worse* micronutrient deficiencies under no-tillage. In only one case has a response to zinc under no-tillage been obtained, and this was on a field where no residue was left (all corn stubble having been removed for silage) and on which severe erosion had occurred. Therefore, for the moment at least, we feel safe in stating that micronutrient availability under no-tillage is less of a problem than it is under conventional tillage. Future work may shed further light on our understanding, however, and there is no final verdict in at this time.

Role of Cover Crops in Nitrogen Fertilization in No-Tillage

Cover Crops as Green Manure. According to a translation (Wedderbuan and Collingwood, 1876), Xenophon (434–355 B.C.) wrote that weeds plowed into the soil "will afford a ready manure for the soil." In recommending spring plowing, he suggested that weeds are large enough at that season to serve as manure, but not having shed seed, they will not spring up again. This is probably the earliest record of using green manure. The superiority of legumes as green manure crops was recognized later. According to Semple (1928), Theophrastus (372–287 B.C.) recorded the use of bean crops as green manure by farmers in Macedonia and Thessaly. Semple also stated that other ancient writers (Cato, 234–149 B.C.; Columella, about 45 A.D.) rated legumes according to their suitability for improving the soil; lupine was generally considered to be the best legume for green manure.

The practice of green manuring was probably lost during the Middle Ages but was renewed during the beginning of the era of modern agriculture. Lawes and Gilbert at Rothamsted, England, about 1855, concluded from their experiments that legume crops (clover) can grow well without much nitrogen in the soil and presumed that legumes could in some way use nitrogen from the air (Russell, 1966).

In association with appropriate *Rhizobium* bacteria, legume plants are capable of fixing atmospheric nitrogen (N_2) into organic combinations. The bacteria penetrate the root hairs of legume plants, form nodules, and fix nitrogen from the air for the use of both the bacteria and the legume plants. The nitrogen becomes available to other plants through death and decomposition of the legume plants or parts thereof.

Potential uses of legumes to provide nitrogen in nonlegume cropping systems include winter annual legume cover crops for summer nonlegumes, summer annual legumes to produce and store nitrogen for nonlegumes grown in the fall or the next summer, legume crops in rotation with nonlegume crops and biennial or perennial legumes grown in association with nonlegumes. Traditionally, legume plants have been used in rotations and as winter cover crops or as components of a mixture of forage species for hay and pasture. In row crop production, the legumes were usually plowed under in conventional tillage systems and a nonlegume crop planted to use the biologically fixed nitrogen. In addition, a cover crop increases soil organic matter beyond the amount contributed by crop residue (Moschler *et al.,* 1975), providing a number of indirect benefits.

According to Allison (1957), average values for nitrogen fixation by legume crops are usually in the range of 56–112 kg/ha but 225 kg/ha or more nitrogen may be fixed by certain legumes, if inoculated with the appropriate strain of *Rhizobium*. Species of legumes differ inherently in their proficiency in fixing nitrogen. Alfalfa is generally regarded as highest, with the clovers next. Allison (1957) stated that the amount of nitrogen fixed by a particular legume species in a nitrogen deficient soil is dependent upon the dry matter production of the crop. Therefore, any factor limiting dry matter production by the legume decreases the amount of nitrogen fixed. Triplett, Haghiri and Van Doren (1979) observed that it did not matter that the legume stand was weedy as long as there was a reasonable population of vigorous legume plants in the stand. Legumes which are poorly adapted to the specific locality will perform poorly in fixing nitrogen and providing it to the subsequent crop. Also, the ratio of the amount of nitrogen in the above-ground portion to that of the roots depends on the top-root ratios of the legumes. In winter annual legumes, that ratio appears to be about 4:1 (Mitchell and Teel, 1977), but it may be lower for perennial legumes, with as much as one-third of the fixed nitrogen in the roots (Wilkinson and Langdale, 1974).

During the years of inexpensive chemical nitrogen fertilizer, dependence on legumes for nitrogen to support nonlegume crops waned to virtually nonexistent. Because of this, most of our present techniques in the use of legumes are based on a low level of technology, old varieties, and low plant populations. With rapidly escalating nitrogen fertilizer prices and the need for a cover crop to provide soil erosion control, interest in the use of legumes has revived. Management techniques for using legume cover crops in modern cropping and tillage systems, such as reduced tillage and no-tillage, are just now being learned.

Effects of Cover Crops on Nitrogen Efficiency. Along with other benefits, cover crops add nitrogen to the soil (if a legume) and conserve nitrogen which might otherwise be lost by leaching or denitrification. These benefits were clearly demonstrated by results obtained by Adams *et al.* (1970) on a Cecil soil in the Southern Piedmont Region with rye and vetch plowed under each spring. The seven-year average corn grain yields with no nitrogen fertilizer and rye cover crop was more than double the yields without a cover crop. With nitrogen fertilizer treatments of 45, 90 and 180 kg/ha nitrogen, the same trend persisted, although at a lower magnitude than without nitrogen fertilizer. This suggested benefits other than nitrogen conservation. Where vetch was turned under, corn grain yields were about 5,300–5,500 kg/ha regardless of nitrogen fertilizer rate. That compared very favorably to yields of about 1,600 and 3,900 kg/ha, respectively, for treatments of no cover crop and rye cover crop without nitrogen fertilizer. In their experiment, vetch gave no grain yield advantage over rye where nitrogen fertilizer was applied. The other benefits accrued from the cover crops probably were associated with organic matter additions and perhaps protection from soil erosion.

In a study in North Carolina, Welch, Nelson and Krantz (1951) found that vetch provided adequate nitrogen for corn and cotton when plowed under. The vetch without nitrogen fertilizer increased corn yields by 3,260 kg/ha, seed cotton by 468 kg/ha and peanuts by 250 kg/ha compared to no cover crop and no nitrogen fertilizer.

Nitrogen immobilization occurs when there is a net build-up of organic matter with a high C:N ratio, such as crop residue or certain nonlegume cover crops. Several studies show that fertilizer nitrogen is immobilized to a greater extent with no-tillage than with conventional tillage. In addition to the likelihood of greater losses of nitrogen from no-tillage by leaching and denitrification and decreased availability by immobilization, there is a slower rate of mineralization of organic soil nitrogen. The combination of these effects makes nitrogen management more critical in no-tillage than in conventional tillage systems and, in some cases, has resulted in higher rates of nitrogen fertilizer being recommended for no-tillage.

Cover Crops as a Mulch for No-Tillage. Several of the advantages of no-tillage depend on the presence of a mulch cover on the soil surface. Usually the mulch is provided either by winter cover crops or by residues from the preceding crop. The mulch provides protection of the soil from erosion, increases water infiltration, decreases surface runoff, and lowers evaporation losses. In these ways, it enhances the plant-available water supplying capacity of the soil and is probably the main factor accounting for greater soil water use efficiency with no-tillage in comparison to conventional tillage. When cover crops are grown and killed down to form the mulch, they provide cover in addition to that provided by crop residues. Many times there are insufficient amounts of residues from the preceding crop to provide the desired benefits. The soybean is notable for low residue production. Corn or other crops harvested for silage may leave little residue on the surface of the soil. Growing crops also absorb plant nutrients which might otherwise be lost by leaching during the winter and spring (Jones, 1942). These nutrients are recycled and may be available to the primary crop plants as the cover crop plants decompose and their nutrients are mineralized. The effects of this can be seen in Table 5-15 in the greater yields of corn on plots with a rye cover crop compared to a cover of corn stalk residue, which was thought to be due partly to conservation of nitrogen. Another benefit derived from nutrient conservation by cover crops is a decrease in potential pollution of ground and surface water by nutrients.

The use of a cover crop in the production of corn by the no-tillage method usually involves seeding a winter cover crop in the fall and chemically killing it in the spring when the corn is planted. In most cases, the cover crop consists of one or a mixture of the species of winter annuals adapted to the area. Often these are small grains. Legume cover crops can provide the benefits outlined above and,

Table 5-15. Effect of winter cover crop and nitrogen fertilizer rate on yield of no-tillage corn at Lexington, Kentucky; average of 1977–1981. (From Ebelhar, Frye and Blevins, 1982.)

	N FERTILIZER APPLIED (kg/ha)		
COVER TREATMENT	0	50	100
		Grain Yield (kg/ha)	
Hairy vetch	6,410	6,840	9,040
Big flower vetch	4,190	6,630	6,560
Crimson clover	4,430	5,680	7,360
Rye	4,030	5,720	7,580
Corn stalk residue	3,790	5,230	6,820

in addition, supply biologically fixed nitrogen to meet part or all of the nitrogen requirements of no-tillage corn. Several legumes have been proven adaptable for use as cover crops in no-tillage systems, and many others have potential for such use but have not been evaluated. Annual legumes are the ones that have been most commonly used, but some perennial legumes are showing promise in field experiments at several locations.

Results from 1977–1981 in a study underway in Kentucky shows that winter annual legumes can be grown as cover crops in a continuous no-tillage system of corn production. Winter cover treatments used in the experiment were hairy vetch (*Vicia villosa*), big flower vetch (*Vicia grandiflora* W. Koch var. *Kitailbeliana*), crimson clover (*Trifolium incarnatum*), rye (*Secale cereale*), and corn stalk residue. Corn was planted by the no-tillage method directly into the cover crops and corn residue and the cover crops were chemically killed. Each cover treatment was combined with treatments of 0, 50 and 100 kg/ha nitrogen from ammonium nitrate (NH_4NO_3) fertilizer broadcast on the surface at the time the corn was planted.

Average yields of corn grain over the five years are shown in Table 5-15 for the Lexington location. Hairy vetch resulted in much greater yields of corn grain than any other cover treatment. It produced more dry matter and had a higher percentage of nitrogen than any other cover crop in the annual legume experiment (Table 5-16). Based on yields of corn, hairy vetch provided biologically fixed nitrogen equivalent to about 90–100 kg/ha nitrogen to the corn. There was an increase in corn yields with rye cover crop

Table 5-16. Yield and nitrogen content of cover crops at time of corn planting, Lexington, Kentucky, 1980–1981. (From Ebelhar, Frye and Blevins, 1982.)

COVER TREATMENT	DRY MATTER YIELD (kg/ha)	N CONCENTRATION (%)	TOTAL N (kg/ha)
Hairy vetch	5,060	4.32	219
Big flower vetch	1,880	3.24	61
Crimson clover	2,370	2.34	56
Rye	3,360	1.30	44

compared to corn stalk residue cover. This was probably due to both the conservation of nitrogen by plant uptake during the winter and spring and water conservation due to the mulch during the corn growing season. There was virtually no difference in the results with rye, crimson clover and big flower vetch.

The 1981 results of corn yields in a similar experiment in western Kentucky are shown in Table 5-17. Hairy vetch and big flower vetch cover crops resulted in the greatest increases in yields. Compared to corn stalk residue only, the two vetches approximately tripled grain yields without nitrogen fertilizer applied and almost doubled yields with the 50-kg/ha rate of nitrogen fertilizer. The effect was present with the 100-kg/ha rate but was less dramatic. The performance of big flower vetch was comparable to hairy vetch, each providing approximately 90–100 kg/ha biologically fixed nitrogen to the corn. Crimson clover was superior to rye cover cop but inferior to big flower vetch and hairy vetch in affecting corn yields.

Research in other regions of the U.S. show similar results. Mitchell and Teel (1977) found that mixtures of hairy vetch and small grains or crimson clover and small grains as winter cover crops in Delaware resulted in grain yields of no-tillage corn comparable to yields produced by the application of 112 kg/ha fertilizer nitrogen (Table 5-18). As would be expected, these cover crops provided the greatest amounts

Table 5-17. Effect of winter cover and nitrogen fertilizer rate on yield of corn at Princeton, Kentucky, 1981. (From Frye *et al.*, 1981.)

	N FERTILIZER APPLIED (kg/ha)					
	0		50		100	
COVER TREATMENT	OS[1]	D[1]	OS	D	OS	D
	Grain Yield (kg/ha)					
Corn stalk residue	1,760		3,260		6,330	
Rye	1,130	1,820	1,760	3,260	5,210	5,330
Crimson clover	3,200	3,640	4,450	5,020	6,150	5,960
Big flower vetch	4,520	5,580	6,020	5,710	7,340	6,960
Hairy vetch	5,460	5,710	5,960	6,330	7,210	6,770

[1] OS = overseeded before corn harvest; D = drilled in after corn harvested.

Table 5-18. Yield of corn with irrigation in Delaware; average of two years data. (From Mitchell and Teel, 1977.)

COVER CROP	N FERTILIZER (kg/ha)		
	0	56	112
	Grain Yield (kg/ha)		
No cover crop	3,960	6,410	6,980
Spring oats	4,820	6,770	8,420
Spring oats and hairy vetch	8,350	8,780	8,570
Rye	3,960	5,830	7,270
Rye and hairy vetch	6,120	7,200	8,060

of total nitrogen to the soil after being killed before planting corn. An estimated one-third of the total nitrogen produced by the cover crops was released to the corn the first season. The remaining nitrogen would be expected to be released at a slower rate over a period of a few years.

Krenzer (1981) stated that legume cover crops of hairy vetch or crimson clover established during September in a conventionally prepared seedbed provided 67–112 kg/ha available nitrogen to corn or grain sorghum. Work by Tyler (1982) in Tennessee in 1980 and 1981 (Table 5-19) showed that grain yields of no-tillage corn with applications of 0, 56, 112 and 168 kg/ha nitrogen following a mixed winter cover crop of wheat and hairy vetch were 100, 74, 4 and 2 percent, respectively, greater than following a cover crop of wheat

Table 5-19. Effect of cover crops on yield of corn grain at Jackson, Tennessee; average 1980 and 1981. (From Tyler, 1982.)

COVER CROP	TILLAGE	N RATE (kg/ha)			
		0	56	112	168
		Yield (kg/ha)			
None	Conv.	1,500	4,010	5,640	6,080
Wheat-vetch	Conv.	2,700	5,020	5,710	6,080
Wheat-vetch	No-till	1,880	3,830	4,830	5,210
Wheat	No-till	940	2,200	4,640	5,080

alone. The data indicated that the hairy vetch contributed biologically fixed nitrogen to no-tillage corn equivalent to about 112 kg/ha fertilizer nitrogen. Yields were somewhat greater (500–1,200 kg/ha grain) under conventional tillage than no-tillage following the wheat-hairy vetch cover crop. This was probably a reflection of the slower rate of nitrogen mineralization usually observed in no-tillage systems.

Flannery (1981) reported on a three-year study comparing yields of conventional tillage and no-tillage corn silage following rye or hairy vetch cover crops (Figure 5-7). Three, 67-kg/ha increments of fertilizer nitrogen and two rates of potassium fertilizer were used with each cropping treatment. With 200 kg/ha potassium fertilizer (K_2O) and hairy vetch cover crop, there was no response to nitrogen fertilizer rates up to 200 kg/ha nitrogen. Without nitrogen fertilizer applied, corn yields on the hairy vetch plots were about equal to yields on rye cover crop with 200 kg/ha nitrogen fertilizer. Yields

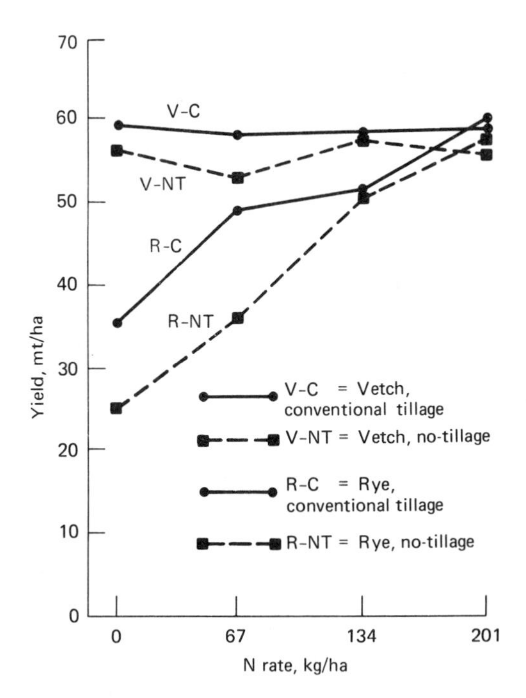

Figure 5-7. Corn Silage yields under no-tillage and conventional tillage following cover crops, hairy vetch or rye. (Reprinted from 1981 Summer-Fall issue of BETTER CROPS, "Conventional versus No-tillage Corn Silage Production," by Ray L. Flannery.)

with the hairy vetch cover crop were the same under both no-tillage and conventional tillage. There was a substantial response to nitrogen rates following rye cover crop. The results indicated that hairy vetch provided between 135 and 200 kg/ha nitrogen to the corn silage.

Hargrove, Touchton, and Duncan (1982) in Georgia planted grain sorghum *(Sorghum bicolor)* by no-tillage into a crimson clover cover crop which yielded 3,768 kg/ha and contained 80 kg/ha nitrogen. Yields of grain sorghum averaged 5,354 kg/ha and were unaffected by nitrogen fertilizer rates ranging from 15–135 kg/ha nitrogen.

Martin and Touchton (1982) reported on work in Alabama with cover treatments of common vetch *(Vicia sativa)*, Austrian winter pea *(Pisum sativa)*, crimson clover, and arrowleaf clover *(Trifolium visculosum)*, rye and winter fallow. Where the legumes were not harvested, they produced about 100 kg/ha nitrogen in the above-ground portions. No-tillage grain sorghum yields were 4,400 kg/ha with no nitrogen fertilizer and were not increased by fertilizer nitrogen rates up to 100 kg/ha. With rye and winter fallow plots, 100 kg/ha fertilizer was required to obtain equal yields of grain sorghum.

A few investigations are underway on the use of perennial legumes as cover crops for no-tillage corn. Perennial legumes have an obvious advantage in that they do not have to be planted each fall as do annual legumes. Another advantage observed in preliminary work in Kentucky is that perennial legumes appear to be more efficient than annual legumes in producing nitrogen for no-tillage corn. Two disadvantages of perennial legumes are that (1) many are slow and expensive to establish a stand and (2) they may be difficult to manage in the no-tillage corn system. Some perennial legumes require from several months to a year to become established well enough to be suitable for planting no-tillage corn. In management, the legumes must be maintained alive but in a severely suppressed state during the corn growing season to prevent serious competition for soil water. Regrowth must occur that fall or early the next spring to produce adequate mulch and provide nitrogen for the next corn crop. Such precision is sometimes difficult to attain.

Preliminary results in 1981 from a perennial legume cover crop experiment in Kentucky indicated that crownvetch *(coronilla varia)* mulch supplied an abundance of nitrogen to no-tillage corn without nitrogen fertilizer (Table 5-20). The crownvetch was mowed at corn

Table 5-20. Dry matter yield and nitrogen content of cover crops at corn planting in a perennial legume cover crop experiment at Lexington, Kentucky, 1981. (From Frye and Blevins, 1981.)

COVER TREATMENT	DRY MATTER YIELD (kg/ha)	N CONCENTRATION (%)	TOTAL N (kg/ha)
Crownvetch	7,120	3.40	242
Alfalfa	3,390	3.10	105
Big flower vetch	4,290	3.50	150

planting and was severely suppressed later with a directed spray of paraquat. Grain yields averaged 8,470 kg/ha without irrigation. In that same experiment, plots where big flower vetch was used as a cover crop produced 7,530 kg/ha corn grain without nitrogen fertilizer. Big flower vetch is an annual legume, but it produces an adequate number of mature seed early enough to reseed itself before it has to be killed to plant no-tillage corn.

An advantage of big flower vetch over crownvetch or other perennials is that its life cycle is completed and regrowth does not occur after it has been mowed or chemically killed at corn planting. This prevents it from competing with the corn for available soil water during the growing season. Weed control must be maintained chemically. The mature seeds which fall to the ground do not germinate until cold weather, at about the time the corn is harvested in the fall, thus beginning the cycle anew.

A possible alternative to maintaining perennial legumes severely suppressed is allowing them to continue growing after suppression by a light application of herbicides or by close mowing and irrigating the corn crop to avoid drastic yield reductions due to water competition. Such intercropping requires less herbicides and lowers the risk of killing the legume stand in the process of suppressing it. This cropping system would produce more biomass and biologically fixed nitrogen in the legumes. The additional biomass yield could be left on the soil to further decrease the nitrogen fertilizer needs for the corn crop or be removed as forage for livestock. The relative effects of the removal of cover crops for forage is discussed below. Hartwig (1974) in Pennsylvania reported that yields of corn with two irrigations during the season averaged 10,350 kg/ha with 105 kg/ha fertilizer nitrogen on plots of crownvetch lightly suppressed by herbicides. The grain yields ranged up to 13,170 kg/ha in those tests. Hartwig recommended establishing crownvetch as a perennial cover crop for no-tillage corn by seeding 5.6 kg/ha when planting corn or soybeans. The crownvetch may be seeded broadcast or banded over the corn or soybean rows using a granular insecticide or herbicide box attached to the planter. The same seeding rate is recommended regardless of seeding method.

Effects of Removal for Forage. Work by Wilkinson and Dobson (1981) in Georgia shown in Table 5-21 indicated that removal of the above-ground portion of crimson clover and hairy vetch cover crops before planting corn resulted in a severe decrease in yield of both conventional and no-tillage corn where no nitrogen fertilizer was applied. Where 118 kg/ha fertilizer nitrogen was applied, removal of top growth had little effect on yield of conventional tillage corn but substantially decreased yields in no-tillage corn. Considerable effectiveness of legume cover crops in providing nitrogen to corn was lost when the top growth was removed as forage. The range of decreased effectiveness was from 20 percent with conventional tillage to 30 percent with no-tillage. The greater effect on no-tillage corn from the removal of top growth was probably a result of the slower rate of nitrogen mineralization generally associated with no-tillage.

Table 5-21. Effect of removal of cover crops on yield of corn grain in Georgia, 1976–1977. (From Wilkinson and Dobson, 1981.)

		CRIMSON CLOVER		HAIRY VETCH	
N RATE (kg/ha)	COVER CROP MANAGEMENT	CONVENTIONAL	NO-TILL	CONVENTIONAL	NO-TILL
		Corn Yields (kg/ha)			
0	Removed	4,080	5,710	5,770	4,520
0	Mulch left	5,080	8,150	7,900	6,330
118	Removed	8,220	6,900	7,780	6,710
118	Mulch left	7,780	8,340	8,280	7,780

Martin and Touchton (1982) compared the effects of grazing and cutting for hay with accumulation of winter growth of four cover crops on the nitrogen content of the legumes and yield of grain sorghum. The nitrogen content of the above-ground portions of the legumes averaged 45, 40 and 100 kg/ha for grazed, hayed and accumulated plots, respectively. Grain sorghum yields on the grazed and hayed treatments with 34 kg/ha fertilizer nitrogen yielded the same as on the accumulated treatments with no nitrogen fertilizer. Thus, it appears that removal of most of the top growth decreased the amount of nitrogen by more than one-half and decreased the amount of nitrogen supplied by the legume cover crops to no-tillage grain sorghum by about 34 kg/ha.

Legume Cover Crops in No-Tillage Vegetable Production. In 1982 an experiment was established at the University of Kentucky at Lexington in which vegetable crops were grown by no-tillage and conventional tillage in vetch cover crops, killed to form a mulch (Knavel, 1982). Spring-summer crops of sweetcorn, snap bean, summer squash, tomato and cabbage are planted into hairy vetch killed with glyphosate (Round-up) in late April. Fall crops of cabbage, broccoli, and cauliflower are planted into a killed cover crop of American joint vetch (*Vicia americana* Muhl.) in early August. Nitrogen fertilizer is applied at the rates of 56 and 112 kg/ha N, one-half at planting and one-half as a sidedress treatment. Check plots have legume cover crops but no nitrogen fertilizer applied. No-tillage production is compared with production by conventional tillage to determine the amount of nitrogen provided by the vetches and to determine their effects on yield and quality of the vegetables. Since 1982 is the first year of this experiment, yield and quality data have not been measured at the time of this writing.

REFERENCES

Adams, W. E., H. D. Morris, and R. N. Dawson. 1970. Effect of cropping systems and nitrogen levels on corn *(Zea mays)* yields in the Southern Piedmont Region. *Agron. J.* **62**:655–659.

Allison, F. E. 1957. Nitrogen and soil fertility. *Soil: The 1957 Yearbook of Agriculture.* Government Printing Office, Washington, D.C.

Bandel, V. A. 1981. Personal communication. Department of Agronomy, Maryland Agric. Exp. Sta., College Park, Maryland.

Belcher, C. R. and J. L. Ragland. 1972. Phosphorus absorption by sod-planted corn (*Zea mays* L.) from surface-applied phosphorus. *Agron. J.* **64**:754–756.

Best, J. A., J. B. Weber, and T. J. Monaco. 1975. Influence of soil pH on S-triazine availability to plants. *Weed Sci.* **23**:378–382.

Blevins, R. L., G. W. Thomas, and P. L. Cornelius. 1977. Influence of no-tillage and nitrogen fertilization on certain soil properties after 5 years of continuous corn. *Agron. J.* **69**:383–386.

Blevins, R. L., G. W. Thomas, M. S. Smith, W. W. Frye, and P. L. Cornelius. 1983. Changes in soil properties with long-term continuous no-tillage and conventionally tilled corn. *Soil and Tillage Res.* **3**:135–146.

Blevins, R. L., L. W. Murdock, and G. W. Thomas. 1978. Effect of lime application on no-tillage and conventionally tilled corn. *Agron. J.* **70**:322–326.

Doran, J. W. 1980. Soil microbial and biochemical changes associated with reduced tillage. *Soil Sci. Soc. Am. J.* **44**:765–771.

Ebelhar, W., W. W. Frye, and R. L. Blevins. 1982. Unpublished data. Department of Agronomy, University of Kentucky, Lexington.

Flannery, R. L. 1981. Conventional vs. no-tillage corn silage production. *Better Crops* **LXVI**: 3–6. (Summer-Fall).

Frye, W. W. and R. L. Blevins. 1981. Unpublished data. Department of Agronomy, Kentucky Agric. Exp. Sta., Lexington, Kentucky.

Frye, W. W., J. H. Herbek, and R. L. Blevins. 1981. Unpublished data. Department of Agronomy, Kentucky Agric. Exp. Sta., Lexington, Kentucky.

Frye, W. W., G. W. Thomas, K. L. Wells, and M. J. Bitzer. 1979. Unpublished data. Department of Agronomy, Kentucky Agric. Exp. Sta. Lexington, Kentucky.

Hargrove, W. L., J. T. Reid, J. T. Touchton, and R. N. Gallaher. 1982. Influence of tillage practices on the fertility status of an acid soil double-cropped to wheat and soybeans. *Agron. J.* **74**:684–687.

Hargrove, W. L., J. T. Touchton, and R. R. Duncan. 1982. Nutrient content of crimson clover forage and its influence on grain sorghum production under no-tillage management. *Abstracts of Technical Papers, No. 9.* Southern Branch Am. Soc. Agron., Orlando, Florida, Feb. 7–10.

Hargrove, W. L. and G. W. Thomas. 1981. Effect of organic matter on exchangeable aluminum and plant growth in acid soils. Chapter 8 in *Chemistry in the Soil Environment.* Am. Soc. Agron., Madison, Wisconsin, pp. 151–166.

Hartwig, N. L. 1974. Crownvetch makes a good sod for no-till corn. *Crops and Soils* **27**(*3*):16–17.

Hill, J. D. and R. L. Blevins. 1973. Quantitative soil moisture use in corn grown under conventional and no-tillage methods. *Agron. J.* **65**:945–949.

Jones, R. J. 1942. Nitrogen losses from Alabama soils in lysimeters as influenced by various systems of green manure crop management. *J. Am. Soc. Agron.* **34**:574–585.

Jones, U. S., M. G. Hamilton, and J. B. Pitner. 1979. Atmospheric sulfur as related to fertility of Ultisols and Entisols in South Carolina. *Soil Sci. Soc. Am. J.* **43**:1169–1171.

Jordan, H. V., C. E. Bardsley, L. E. Ensminger, and J. A. Lutz. 1959. Sulfur content of rainwater and atmosphere in southern states as related to crop needs. *U.S.D.A. Tech. Bul. 1196.*

Kells, J. J., R. L. Blevins, C. E. Rieck, and W. M. Muir. 1980. Effect of pH, nitrogen and tillage on weed control and corn yield. *Weed Sci.* **28**:719-722.

Kitur, B. K. 1982. Effect of lime particle size and mixing on pH changes in a Maury silt loam (unpublished data). Special problem in Soil Chemistry, University of Kentucky.

Knavel, D. E. 1982. Personal communication. Kentucky Agric. Exp. Sta., Lexington, Kentucky.

Koehler, F. E., M. E. Fischer, and R. W. Meyer. 1977. Fertilization practices for no-till wheat in the Pacific Northwest. *Agron. Abstr.*, p. 161 (Am. Soc. Agron.).

Krenzer, E. G., Jr. 1981. Cover crops and cover crop management. Paper presented at Southeastern No-till Systems Conf. Nov. 24, 1981. North Carolina State University, Raleigh, North Carolina.

Mahtab, S. K., C. L. Godfrey, A. R. Swoboda, and G. W. Thomas. 1971. Phosphorus diffusion in soils: I. The effect of applied P, clay content and water content. *Soil Sci. Soc. Am. Proc.* **35**:393-397.

Martin, G. W. and J. T. Touchton. 1982. Management of winter legumes for grain sorghum production. *Abstracts of Technical Papers, No. 9.* Southern Branch Am. Soc. Agron., Orlando, Florida, Feb. 7-10.

Miller, H., K. Wells, M. Bitzer, G. W. Thomas, and R. E. Phillips. 1975. Influence of time of application of nitrogen fertilizer on corn yields. *Agronomy Notes* **8**, *No. 3.* Department of Agronomy, Kentucky Agric. Exp. Sta., Lexington, Kentucky.

Mitchell, W. H. and M. R. Teel. 1977. Winter-annual cover crops for no-tillage corn production. *Agron. J.* **69**:569-573.

Moncrief, J. F. and E. E. Schulte. 1982. The effect of tillage and fertilizer source and placement on nutrient availability to corn. Iowa's 34th Fertilizer and Ag. Chem. Dealers Conf., Iowa State University, Ames, Iowa.

Moschler, W. W. 1971. Unpublished data. Department of Agronomy, Virginia Polytechnic Institute and State University, Blacksburg, Virginia.

Moschler, W. W., D. C. Martens, and G. W. Shear. 1975. Residual fertility in soil continuously field cropped to corn by conventional and no-tillage methods. *Agron. J.* **67**:45-48.

Olson, S. R. and F. S. Watanabe. 1963. Diffusion of phosphorus as related to soil texture and plant uptake. *Soil Sci. Soc. Am. Proc.* **27**:648-653.

Phillips, R. E., C. R. Belcher, R. L. Blevins, H. F. Miller, G. W. Thomas, T. Na Nagara, V. Quisenberry, and M. McMahon. 1971. *Agron. Research Misc. Publ. 394,* p. 38. Kentucky Agric. Exp. Sta., Lexington, Kentucky.

Phillips, R. E., R. L. Blevins, G. W. Thomas, W. W. Frye, and S. H. Phillips. 1980. No-tillage agriculture. *Science* **208**:1108-1113.

Rice, C. W. and M. S, Smith. 1982. Denitrification in no-till and plowed soils. *Soil Sci. Soc. Am. J.* **46**:1168-1173.

Russell, E. J. 1966. *A History of Agricultural Science in Great Britain, 1620–1954.* George Allen and Unwin, Ltd., London.

Semple, E. C. 1928. Ancient Mediterranean agriculture. Part II. Manuring and seed selection. *Agric. Hist.* **2**:129–156.

Shear, G. M. and W. W. Moschler. 1969. Continuous corn by the no-tillage and conventional tillage methods: A six-year comparison. *Agron. J.* **61**:524–526.

Shenk, M. and J. Saunders. 1980. Personal communication. Centro Agronomico Tropical De Investigacion Y Ensenaza, CATIE, Programa de Plantas Perennes, Turrialba, Costa Rica.

Singh, T. A., G. W. Thomas, W. W. Moschler, and D. C. Martens. 1966. Phosphorus uptake by corn (*Zea mays* L.) under no-tillage and conventional practices. *Agron. J.* **59**:147–148.

Thomas, G. W. 1975. The relationship between organic matter content and exchangeable aluminum in acid soil. *Soil Sci. Soc. Am. Proc.* **39**:591.

Thomas, G. W., R. L. Blevins, R. E. Phillips, and M. A. McMahon. 1973. Effect of a killed sod mulch on nitrate movement and corn yield. *Agron. J.* **65**:736–739.

Touchton, J. T. 1982. Personal communication. Department of Agronomy and Soils. Alabama Agric. Exp. Sta., Auburn University, Auburn, Alabama.

Triplett, G. B., Jr. and D. M. VanDoren, Jr. 1969. Nitrogen, phosphorus and potassium fertilization of non-tilled maize. *Agron. J.* **61**:637–639.

Triplett, G. B., Jr., F. Haghiri, and D. M. VanDoren, Jr. 1979. Legumes supply nitrogen for no-tillage corn. *Ohio Report on Res. and Dev.* **64**:83–85 (Nov.-Dec.).

Tyler, D. D. 1982. Unpublished data. Tennessee Agr. Exp. Sta. at the West Tennessee Exp. Sta., Jackson, Tennessee.

Tyler, D. D. and G. W. Thomas. 1977. Lysimeter measurements of nitrate and chloride losses from soil under conventional and no-tillage corn. *J. Environ. Qual.* **6**:63–66.

Van Bavel, C. H. M. and J. H. Lillard. 1957. Agricultural drought in Virginia. Virginia Agric. Exp. Sta. Tech. Bul. 128.

Wedderbuan, A. D. O. and W. G. Collingwood. 1876. The economist of Xenophon (English translation). Lenox Hill Pub. and Dist. Co. (Burt Franklin), New York.

Welch, C. D., W. L. Nelson, and B. A. Krantz. 1951. Effects of winter cover crops on soil properties and yields in a cotton-corn and in a cotton-peanut rotation. *Soil Sci. Soc. Am. Proc.* **15**:229–234.

Wilkinson, S. R. and J. W. Dobson, Jr. 1981. Unpublished data. Department of Agronomy, Georgia Agric. Exp. Sta., Athens, Georgia.

Wilkinson, S. R. and G. W. Langdale. 1974. Manures and legumes as alternate nitrogen sources for crop production. *Proceedings of Fertility Workshop.* Rural Development Center, Tifton, Georgia, Oct. 30–Nov. 1, 1974, pp. 48–55.

6
Energy Requirement in No-Tillage

Wilbur W. Frye
Associate Professor of Agronomy
University of Kentucky

ENERGY USE IN AGRICULTURE

The production phase of U.S. agriculture uses large amounts of fossil energy as gasoline, diesel fuel, natural and L-P gas, oil, electricity, fertilizers, pesticides, feeds, seeds and machinery. Figure 6-1 shows an estimated division of the energy among the major uses in production agriculture. About one-third of the energy is directly from fossil fuels. Indirect inputs of fossil energy as fertilizers, pesticides, feeds, seeds machinery and electricity make up the remaining two-thirds. Of the fuels, gasoline comprises about 40 percent, diesel fuel 32 percent, natural gas 16 percent and L-P gas 12 percent (USDA-ERS-FEA, 1977). Most of the on-farm use of diesel fuel and gasoline is in farm tractors, trucks and automobiles, uses for which energy substitution is for the most part impractical. L-P gas is used mainly for on-farm crop drying. The major on-farm use of natural gas is for irrigation and for grain drying, especially in large commercial operations. Much of the indirect use of natural gas in agriculture is in the form of nitrogen fertilizer, since almost all nitrogen fertilizer used in the U.S. is manufactured from natural gas. An estimated 89 percent of the energy used in production agriculture goes for crop production and the remaining 11 percent is used for livestock production (Soil Conservation Society of America, 1978).

Although a large amount of fossil energy is used, U.S. agriculture has a history of a high level of efficiency in terms of the ratio of production output to energy use (Steinhart and Steinhart, 1974). As

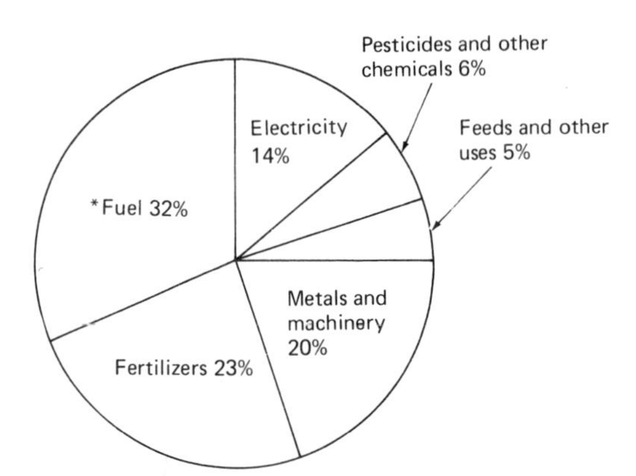

Figure 6-1. Distribution of energy used in production agriculture in U.S.A. (From American Chemical Society, 1974.) *Values are percent of energy used in production agriculture.

in many other areas of our society, however, agricultural output per unit of energy input began to level off somewhat in recent years as the substitution of fossil energy became excessive. In some cases, oversized tractors and other machinery, excessive use of nitrogen fertilizer, extravagant substitution of fossil energy for human energy and other forms of energy waste have undoubtedly contributed to the decline in energy-use efficiency. The ratio is still quite favorable, but the decline in efficiency points out a need and an opportunity for energy conservation and improved efficiency in production agriculture. Management practices that produce higher yields more efficiently can improve energy-use efficiency although greater inputs of energy may be required. Emphasis should be on conservation through improved efficiency. The result will be lower overall energy demand.

In addition to using energy efficiently, agriculture produces large amounts of chemically combined energy through photosynthesis. Energy from the sun is combined in the form of agricultural products – food, feed, oil, fiber, crops harvested for energy, wood and wood products and crop residues. In the case of most crops, energy in the products exceeds the production energy. Crops listed by Heichel (1976) as ones that produce more food energy than they require in production energy include all of the grain crops, sugarcane, sugarbeet, peanut, potato, apple and grape. Corn grain, corn silage, sorghum and sugarcane are the more efficient ones, with ratios of energy output to energy input in the range of 4 or 5 to 1. Wheat, oats and soybeans are about 2.5 or 3 to 1, and the others listed are lower. On the other hand, production energy exceeds the energy in the products in the case of some crops. This is more likely to be true of the specialty crops which require intensive culture, including many of the fruit and vegetable crops (Heichel, 1976). These are mostly crops that add quality to the human diet, perhaps justifying the high cost of energy to produce them.

To be a suitable source of energy, a commodity or process must produce more useful energy than it consumes. The fact that energy of the sun can be chemically combined in organic compounds very efficiently in many crops forms the basis for biomass energy from crop residues, alcohols and other plant materials. It gives agriculture the opportunity to be a net producer of energy and the potential to possibly become energy-independent in the future. The sun is a source of free energy, but capturing and storing the sun's energy may be costly in terms of the investment of capital, technology, fossil energy and other resources. At present, photosynthetic efficiency is in the range of 1–3 percent, but is theoretically about 12 percent (National Research Council, 1977). Genetic and other kinds of improvements in photosynthetic efficiency could improve the ratio of output to input energy and, as pointed out by the National Research Council, could be the lowest-cost means of increasing the use of the energy of the sun. Such developments would hasten the time at which fuels from biomass can favorably compete with fossil fuels throughout our economy. In its statement on energy and conservation of natural resources, the Soil Conservation Society of American (1978) pointed out the value of crop residues in maintaining soil organic matter content, controlling soil erosion and feeding livestock, and recommended that crop residues not be used as a source of biomass energy. The society recommended that crops or trees be grown expressly for use in producing biomass energy. Table 6-1 shows the estimated energy content of the harvested portion of several common field crops.

As it pertains to tillage, the main concern is for conservation of energy in crop production. The concept of conservation is mainly one of efficiency in the use of energy and not necessarily lower energy use. The remainder of this chapter discusses energy uses in

Table 6-1. Estimated energy content in harvested portion of some common field crops. (Source: July/August 1981 issue of *Solutions Magazine* [Murphy, 1981].)

CROP	BIOMASS ENERGY (MJ/kg)
Alfalfa	16.3
Corn	15.6
Grain sorghum	16.0
Soybeans	23.6
Wheat	18.4

crop production and points out ways in which energy can be conserved by more efficient use. Emphasis is on those areas which differ substantially between tillage systems.

NEED FOR ENERGY CONSERVATION IN AGRICULTURE

The total food system uses about 16.5 percent of the U.S. energy demand, but only about 3 percent of the energy use in the U.S. goes into production agriculture. In 1981, that amounted to approximately 2.3 trillion MJ used for on-farm production of food out of the estimated 77.9 trillion MJ total U.S. energy demand that year (Energy Information Center, 1982). As the need for agricultural production increases, greater amounts of petroleum fuels will be demanded, especially in developing countries. Worldwide, the increase is expected to be 77 percent during the 13-year period between 1972–73 and 1985–86 (Stout, 1979). The increase for North America during the same period is estimated at about 38 percent. By far the greatest increases in energy use for food production in the near future are expected to be in the developing countries of Africa, Latin America, the Near East and the Far East. Overall, the expected increases average approximately 210 percent in those countries during the 13-year period. The supply of fossil fuels is finite and is predicted to become severely limited during the next 20–30 years (Soil Conservation Society of America, 1978). As the supply decreases, the price will increase, making conservation an important feature of energy use in all areas of our society. Generally, it is more economical to save energy than to increase its production.

Because a small proportion of the total U.S. energy is used in production agriculture, continued expansion of energy inputs into agriculture would have a rather small effect on overall energy use. Furthermore, the effects of energy conservation by farmers on the total energy picture will be small; however, energy conservation in farming is important for three reasons: (1) to save on production costs and increase profit, or place farmers in a more competitive position; (2) to help extend the useful supply of fossil fuels, on which U.S. agriculture is so greatly dependent and for which substitutions are presently impractical; and (3) to foster an attitude of energy conservation in all segments of our society. D'Arge (1981) has stated that "few public policy issues are as pressing as the impact of higher energy costs on the productivity of American agriculture." He further said that rapidly increasing energy prices will provide incentives toward efficiency in the use of energy. Perhaps in no other industry does economics control more than in agriculture. With farm units in the U.S. numbering about two million, farmers cannot readily pass higher costs along to consumers. This situation results in incentives to improve efficiency. To be most effective, energy conservation must be directed at the areas of greatest use. It is essential that management practices aimed at conservation of energy not decrease production per unit. If so, any economic advantage gained by energy conservation would be quickly lost in lower profits. With world population increasing and the demand for food increasing along with it, conservation measures which result in an overall decrease in food production would be an unreasonable alternative. Also, with the present relationship between labor and energy costs, energy conservation practices must not substantially increase labor if they are to be profitable. To be adopted by farmers, practices must be both practical and economical.

ENERGY USE AND EFFICIENCY IN TILLAGE SYSTEMS

Table 6-2 is a list of management inputs and operations often used in crop production and the estimated energy use associated with each. Energy estimates vary greatly among sources in the literature,

Table 6-2. Estimated average energy requirements of crop production inputs and operations.

MANAGEMENT INPUT OR OPERATION	UNIT	DFE[1] (LITERS/UNIT)
Machinery manufacture and repair	kg machinery	2
Primary tillage		
Moldboard plow (20-cm depth)	ha	17
Chisel plow (20-cm depth)	ha	11
Disk (once)	ha	6
Secondary tillage		
Disk	ha	6
Pulvamulcher (after moldboard plow)	ha	6
Spike-tooth harrow	ha	3
Field cultivator	ha	6
Subsoiler (35-cm depth)	ha	20
Plant (90-cm rows)		
Conventional and reduced tillage	ha	4
No-tillage	ha	5
Weed control		
Herbicides	kg a.i.	4
Spray herbicides	ha	1
Apply herbicides and disk second time	ha	7
Cultivate (each time)	ha	4
Fertilizer		
Nitrogen	kg N	1.4
Phosphorus	kg P_2O_5	0.2
Potassium	kg K_2O	0.1
Broadcast granular fertilizer	ha	2
Spray liquid fertilizer	ha	2
Apply anhydrous ammonia (no-tillage)	ha	11
Apply anhydrous ammonia (plowed soil)	ha	7
Irrigation	ha	289
Harvest		
Corn picker-sheller	ha	13
Combine	ha	15
Dry grain (from 23 to 15 percent)	ton	15
Miscellaneous		
Shred cornstalks	ha	7
Disk cornstalks	ha	4
Grain drill	ha	5
Seed	kg	0.4

[1] Diesel fuel equivalent (41 MJ/liter).

and the amount of energy used to represent a particular input or operation depends on the source of the estimate. The estimates made in Table 6-2 are intermediate values based on several sources.

In crop production, the greatest use of energy is for tillage and nitrogen fertilizer. This is especially true in the production of nonirrigated crops which are not dried artificially. When used intensively, either irrigation or crop drying may be the largest input of energy; however, these uses are affected little by tillage systems and will not be discussed in this chapter.

Tillage Operation

The amount of energy required in producing crops by various tillage systems is generally proportional to the amount of tillage involved. This is particularly true if the same rate of nitrogen fertilizer is used with each tillage system. Of the commonly used primary tillage implements, the moldboard plow requires the most energy, the chisel plow is intermediate, and the disk requires the least amount. Conventional tillage, which involves moldboard plowing as the primary tillage and disking at least once as secondary tillage to prepare a seedbed, requires a relatively large amount of fuel. Sometimes cultivation is used in the conventional tillage system for weed control after the crop is established. Reduced tillage systems usually involve chisel plowing or disking as primary tillage and seedbed preparation by secondary tillage with a disk or a field cultivator. Thus, reduced tillage operations require less fuel energy than conventional tillage because less tillage is done. No-tillage eliminates both primary and secondary tillage (including cultivation) and has the lowest energy requirement. Chemical herbicides are used for weed control in no-tillage as well as in most cases with reduced and conventional tillage.

Planting

Most indications are that planting in untilled soil requires slightly more fuel than planting in a conventionally prepared seedbed. Comparisons of no-tillage and conventional tillage show that soil bulk density values are greater for no-tillage at planting time. Thus it seems reasonable that more energy would be required to pull the no-tillage planter at the proper seeding depth through the firmer soil. However, that is not always true. Collins, Williams and Kemble (1980) found that no-tillage planting required the least amount of energy on a Sassafras sandy loam soil in Delaware. The more the soil was tilled before planting, the more energy required to plant the crop. They attributed this to the excessive looseness of the sandy soil.

Field Machinery

One of the justifications often pointed out for the use of large tractors and field machinery is the need for timeliness in order to take maximum advantage of short periods of favorable weather in temperate regions (Wittwer, 1981). Intensive mechanization with large equipment allows growers to conduct farming operations rapidly and efficiently in terms of production output per farm worker. It is estimated that, on the average, one farm worker produces enough food for 60 persons; however, to accomplish this, crop production has become energy-intensive and energy-dependent. Reduced tillage decreases the time needed to till and plant crops, thereby decreasing the need for large, time-efficient equipment and allowing emphasis to be placed on energy-efficiency with regard to equipment sizes.

Another factor of some importance is that less machinery is needed for reduced tillage, especially no-tillage. If continuous no-tillage is practiced, a moldboard plow and disk are not needed, whereas they would be needed in conventional tillage. Phillips *et al.* (1980) estimated the annual machinery manufacturing and maintenance requirements for no-tillage and conventional tillage on a 240-ha farm at 9.5 and 11.6 kg/ha, respectively. According to the estimate of Wittmuss, Olson and Lane (1975), the energy required to manufacture and maintain farm machinery is equivalent to about 2.1 liters of diesel fuel per kg of machinery. Using these estimates, no-tillage requires approximately 18 percent less energy than conventional tillage for manufacturing and maintaining machinery.

It has been suggested by some writers that no-tillage does little to decrease the amount of machinery purchased, because, it is argued, most no-tillage farmers will have the usual complement of tillage implements and will practice conventional tillage every few years. Even if this is true, the machinery (including the tractor) need not be as big and may last longer, resulting in energy saved.

Seeding Rates

Because of the potential for decreased stands with less tillage, higher seeding rates are recommended as tillage is reduced. It is generally recommended that about 20 percent more seeds be planted with no-tillage than with conventional tillage. Other forms of reduced tillage probably require fewer seeds/ha than no-tillage but more than conventional tillage. The higher seeding rate recommended for no-tillage corn over conventional tillage is estimated to be equivalent to about 38 MJ/ha of energy (Phillips *et al.*, 1980) or about 0.9 liter/ha diesel fuel equivalent (DFE).

Weed Control

Estimates of the energy values of herbicides vary widely. The widest variations among materials is probably due mainly to the fact that petroleum or a petroleum product is the carrier for some herbicides. Thus, both the active ingredient and the carrier represent energy. Materials differ in amounts of active ingredients. Finally, different authors use different energy values for the same herbicides. These differences do not appear to be a serious problem in making energy estimates in tillage systems because herbicides represent such a small part of the total energy in the system. In fact, it seems practical to arrive at a single value that can be used to represent all or nearly all herbicides. The energy value used in Table 6-1 is thought to be fairly representative for most herbicides commonly used in crop production, except for paraquat (1,1′-dimethyl-4,4′-bipyridinium ion). It is estimated to represent energy equivalent to approximately 10 liters of diesel fuel per kg of active ingredient (a.i.).

As tillage is decreased, the requirements for pesticides increase, especially herbicides. Kentucky recommendations indicate that about 50 percent more herbicides-insecticides are required in no-tillage than in conventional tillage. According to estimates by Phillips *et al.*

(1980), this amounts to about 109 MJ/ha more energy for pesticides for no-tillage than for conventional tillage. Pesticides requirements in reduced tillage systems would be more than in conventional tillage but less than in no-tillage. The greater pesticide requirement for reduced tillage and no-tillage offsets some but not all of the energy saved by less tillage.

Comparison of Tillage System

Considering only tillage and planting operations and machinery inputs of energy, reduced tillage and no-tillage systems may require substantially less energy than conventional tillage. According to estimates by Frye and Phillips (1980), energy requirements for the tillage systems of chisel plow, disk and no-tillage were 91, 75 and 54 percent of that for conventional tillage, respectively (Table 6-3). The estimates are in rather good agreement with the estimates of others (Collins, Williams and Kemble, 1980; Griffith and Parsons, 1980; Parsons, 1980; Shelton, 1980). Adding on the small amount of

Table 6-3. Estimated energy requirements for several field operations and inputs in four tillage systems. (From Frye and Phillips, 1980.)

INPUT OF OPERATION	TILLAGE SYSTEM			
	CONVENTIONAL TILLAGE	CHISEL PLOW	DISK	NO-TILLAGE
	DFE (liters/ha)			
Moldboard plow	17			
Chisel plow		11		
Disk	6	6	6	
Apply herbicides and disk second time	7	7	7	
Spray herbicides				1
Plant	4	4	4	5
Cultivate (each time)	4	4	4	
Herbicides (manufacturing)	16	19	22	27
Machinery and repair	17	15	12	6
Total	71	66	55	39

energy associated with the higher seeding rate for no-tillage does not appreciably alter the relationships shown in Table 6-3.

Wittmuss and Yazar (1981) reported corn yields with seven tillage systems in Nebraska which showed no statistically significant different (0.05 percent level) in yields of corn among tillage systems, except for one year in which moldboard plowing resulted in decreased yields. A comparison of estimated energy requirements for three of the systems is shown in Table 6-4. Their management treatments included 11.2 kg/ha of insecticide at planting, which was omitted from Table 6-4 because a value for the energy it represents is not available.

Four important observations can be made from Table 6-4. (1) There was lower energy required with reduced and no-tillage, as shown

Table 6-4. Energy analysis of corn under three tillage systems in Nebraska. (Adapted from Wittmuss and Yazar, 1981.)

OPERATION OR INPUT	TILLAGE SYSTEM		
	MOLDBOARD PLOW	CHISEL PLOW	NO-TILLAGE
		DFE (liters/ha)	
Moldboard plow	17		
Chisel plow		11	
Disk	12	6	
Apply herbicides	1	1	1
Plant corn	4	4	5
Herbicides (manufacturing)	22	22	22
Machinery (manufacturing and repair)	17	15	6
Fertilizer (168 kg/ha N)	238	238	238
Harvest grain	15	15	15
Dry grain	17	80	80
Chop stalks	7	7	7
Production energy (total of above)	411	399	374
Energy in grain[1]	2,030	2,030	2,030
Energy output:input ratio	4.9	5.1	5.4

[1] Based on 16 MJ/kg of grain and 41 MJ/liter of diesel fuel. Grain yields (three-year average) were approximately 5,200 kg/ha for all tillage systems.

by the lower total production energy. (2) Reducing tillage did not decrease grain yields. This has been observed in numerous other experiments over a period of many years. (3) As a result of (1) and (2), the chemical energy produced in the grain per unit of production energy increased in the reduced and no-tillage systems. (4) Nitrogen fertilizer accounted for 58–64 percent of the production energy in these examples. It should be pointed out in that regard that the 168 kg/ha rate of nitrogen used by Wittmuss and Yazar is not excessive under most soil and climatic conditions. Rather, it is generally considered a moderate rate. Observation (4) points out the potential for energy conservation through nitrogen management and serves to emphasize the necessity of considering nitrogen fertilizer in any energy conservation efforts in crop production.

ENERGY EFFICIENCY IN FERTILIZER MANAGEMENT

Opportunities are good for conserving energy through improved efficiency in fertilizer use, particularly nitrogen fertilizer. Nitrogen is a high-energy fertilizer, requiring about 83 percent of the energy used for manufacturing fertilizer. By contrast, phosphorus and potassium represent about 11 and 6 percent, respectively (Nelson, 1975). Each kg of nitrogen applied in crop production represents about 1.4 liters DFE (Table 6-2). Different nitrogen fertilizers represent only slightly different amounts of energy per unit of nitrogen, depending mainly on the manufacturing process and the nitrogen content of the material (Table 6-5). Using anhydrous

Table 6-5. Energy represented by various forms of N fertilizer, including manufacture, transportation and storage. Source: *Solutions Magazine* [Achorn, 1981].)

N MATERIAL	ENERGY (MJ/kg N)	APPROXIMATE DFE[1] (Liters/kg N)
Urea, prilled (46% N)	58	1.4
Nitrogen solution (32% N)	57	1.4
Ammonium nitrate, prilled (33.5% N)	61	1.5
Anhydrous ammonia (83% N)	54	1.3

[1] Based on 41 MJ/liter of diesel fuel.

ammonia (NH_3) instead of the other forms would save a small amount of energy in the fertilizer material but would require more energy to apply the NH_3 because it must be injected into the soil, bringing the total energy to about that of other forms of nitrogen fertilizer. Probably more important, NH_3 is usually significantly lower priced than the others.

When applied at the conservative rate of 140 kg/ha, nitrogen fertilizer alone would account for 200 liters/ha DFE energy investment. This is almost three times the amount of other energy inputs required to bring a crop to the harvest stage using the conventional tillage system as outlined in Table 6-3 and is 5.2 times greater than that required with no-tillage. Estimates by Phillips *et al.* (1980) show that nitrogen fertilizer comprises approximately 58 and 64 percent of the energy required to produce a crop of corn by conventional tillage and no-tillage, respectively, if artificial drying and irrigation are not used.

Because nitrogen fertilizer represents a large amount of energy, it presents one of the best opportunities to conserve energy by (1) improving efficiency in the use of commercial nitrogen fertilizer, (2) using legume crops grown in rotation or in association with nonlegume crops or as cover crops for nonlegumes, and (3) using waste materials as sources of nitrogen.

Tillage and Nitrogen Fertilizer Efficiency

Results from Kentucky, Maryland and Virginia show that nitrogen fertilizer is used more efficiently in no-tillage than conventional tillage corn, when evaluated on the basis of yield response on fertilized plots (Frye *et al.*, 1981; Moschler and Martens, 1975; Phillips *et al.*, 1980; Stanford *et al.*, 1979). Table 6-6 shows results from four locations in Kentucky. These data show that corn grain yields were lower under no-tillage than conventional tillage with no nitrogen fertilizer applied; but, at the 170 kg/ha rate of nitrogen fertilizer, no-tillage corn yielded more than conventional tillage corn. Yields were about equal for the two tillage systems with the 85-kg/ha rate of nitrogen fertilizer.

Table 6-6. Corn yields under conventional tillage and no-tillage with 0, 85 and 170 kg/ha N fertilizer on four soils in Kentucky. (From Frye *et al.*, 1981.)

	N APPLIED (kg/ha)					
	0		85		170	
SOIL	CONVENTIONAL	NO-TILL	CONVENTIONAL	NO-TILL	CONVENTIONAL	NO-TILL
	Yield of Grain (kg/ha)					
Maury	5,958	4,767	8,028	7,715	7,840	8,028
Baxter	9,094	9,972	9,283	11,039	9,283	11,039
Cavode	7,903	5,143	8,154	8,655	8,279	9,784
Monongahela	5,814	3,011	9,063	6,931	9,063	10,587
Average	7,192	5,723	8,632	8,585	8,616	9,860

The energy output:input ratio (Energy O/I) for nitrogen fertilizer can be calculated by the following relationship:

$$\text{Energy } O/I = \frac{(Y_{m+1} - Y_m)(E^P)}{(\mathrm{N}_{m+1} - \mathrm{N}_m)(E^\mathrm{N})} \qquad [1]$$

where

Y_m (kg/ha) = yield of crop of mth increment of applied N fertilizer, $m = 0, 1, 2, 3, \ldots$

E^P (MJ/kg) = amount of energy per kg of crop produced by mth increment of fertilizer

N_m (kg/ha) = amount of mth increment of N fertilizer applied

E^N (MJ/kg) = estimated amount of energy represented by mth increment of N fertilizer.

The equation for the first increment of N fertilizer is

$$\text{Energy } O/I = \frac{(Y_1 - Y_0)(E^P)}{(\mathrm{N}_1 - \mathrm{N}_0)(E^\mathrm{N})} \qquad [2]$$

The equation for the second increment of N fertilizer is

$$\text{Energy } O/I = \frac{(Y_2 - Y_1)(E^P)}{(\mathrm{N}_2 - \mathrm{N}_1)(E^\mathrm{N})} \qquad [3]$$

Table 6-7. Yield efficiency and energy efficiency of nitrogen fertilizer as calculated from equation [1] in conventional tillage and no-tillage production of corn, calculated from averages of the four locations shown in Table 6-6.

N FERTILIZER INCREMENT	N FERTILIZER EFFICIENCY (kg grain/kg N)		ENERGY EFFICIENCY (MJ in grain/MJ in N)	
	CONVENTIONAL	NO-TILL	CONVENTIONAL	NO-TILL
First 85 kg/ha	17	34	4.7	9.3
Second 85 kg/ha	0	15	0	4.1

The ratio for the third increment of N fertilizer would be calculated similarly, using $(Y_3 - Y_2)$ and $(N_3 - N_2)$. Table 6-7 shows the Energy *O*/*I* of nitrogen fertilizer for the yield results shown in Table 6-6. E^P is 16 MJ/kg for corn grain at 15.5% moisture, E^N is 58 MJ/kg for the nitrogen fertilizer and, in this case, $(N_{m+1} - N_m)$ is 85 kg/ha.

The 85-kg/ha rate was energy-efficient in both conventional tillage and no-tillage, producing 4.7 and 9.3 MJ in grain per MJ in nitrogen fertilizer for the two systems, respectively. However, the second 85-kg/ha increment (170-kg/ha rate) resulted in an Energy *O*/*I* of 4.1 with no-tillage but 0 with conventional tillage. As was shown in Table 6-6, the 170-kg/ha rate of nitrogen did not increase corn yields over the 85-kg/ha rate in conventional tillage. The overall Energy *O*/*I* values for nitrogen fertilizer (0–170 kg/ha), using the average yields in Table 6-6, are 2.3 with conventional tillage and 6.7 with no-tillage.

Table 6-8 shows calculations from data of Moschler and Martens (1975) from a study conducted in Virginia. Values for the first two levels of nitrogen fertilizer above the check level were very similar for both conventional tillage and no-tillage. The efficiency values for the third level of nitrogen fertilizer (470 kg/ha N) with no-tillage were considerably higher in the no-tillage than in the conventional tillage system. In fact, the efficiency values with conventional tillage were slightly negative. These results are similar to those found in Kentucky, but the soil was much more responsive to nitrogen fertilizer additions than the Kentucky soils, accounting for the higher efficiency values at high nitrogen fertilizer rates.

Table 6-8. Yield efficiency and energy efficiency of nitrogen fertilizer as calculated from equation [1] in conventional tillage and no-tillage production of corn. (Adapted from Moschler and Martens, 1975.)

TOTAL N FERTILIZER[1]	YIELD OF CORN GRAIN[1]		N FERT. EFFICIENCY[2]		ENERGY EFFICIENCY[2]	
	CONVEN-TIONAL	NO-TILL	CONVEN-TIONAL	NO-TILL	CONVEN-TIONAL	NO-TILL
(kg/ha N)	(mt/ha)		(kg grain/kg N)		(MJ in grain/MJ in N)	
68	10.3	9.5	–	–	–	–
202	15.1	14.5	36	38	9.8	10.3
336	17.5	17.7	18	23	5.0	6.5
470	17.3	20.6	−2	22	−.5	6.0

[1] Check treatment had total of 68 kg/ha N applied. N fertilizer and yield values are 3-year totals.
[2] Based on yields and N fertilizer levels above the 68-kg/ha N treatment.

The most important feature of these calculations is the results comparing conventional tillage and no-tillage with regard to the optimum rate of nitrogen fertilizer. According to these results, more energy in the form of nitrogen fertilizer is required to produce economically optimum yields with no-tillage than with conventional tillage, but the energy produced as grain is greater. Therefore, the energy of nitrogen fertilizer is used more efficiently in no-tillage. These data clearly point out that excessive rates of nitrogen fertilizer are energy-inefficient, as well as economically unsound.

Time of Application and Nitrogen Fertilizer Efficiency

Generally, nitrogen fertilizer is more efficient the nearer the time of application is to the time when the crop begins to take up the nitrogen rapidly. Table 6-9 shows that very little nitrogen is taken up by corn plants during the first month after planting. During this period, there is little need for nitrogen and the nitrogen is more susceptible to loss than in any other stage of corn growth. When nitrogen is applied far in advance of rapid uptake, more of it may be lost from the soil by leaching or denitrification or immobilized into soil organic matter. Rapid uptake begins between 31 and 46 days after planting (Table 6-9). Timing the application of the greatest

Table 6-9. Nitrogen accumulation in the above-ground portion of the corn plant at several times during the growing season; average of four replications. (From Na Nagara and Phillips, 1971.)

	N APPLIED, kg/ha			
	0		170	
DAYS AFTER PLANTING	CONVEN-TIONAL	NO-TILL	CONVEN-TIONAL	NO-TILL
	(kg N in plant/ha)			
31	4	4	4	6
46	53	43	58	63
77	112	63	148	156
108	129	67	209	192
127 (harvest)	143	98	205	215

portion of nitrogen to coincide with the time of greatest plant uptake minimizes the effects of leaching, denitrification and immobilization, and improves efficiency.

Results in Table 6-10 from eight locations in Kentucky show that the efficiency of nitrogen fertilizer was improved by delaying application until four to six weeks after the corn was planted. Almost all of the increased efficiency was in the no-tillage corn, mostly with the 85-kg/ha rate of nitrogen. The response of conventional tillage corn to delayed application was mostly negative and there were more

Table 6-10. Effect of delayed applications of nitrogen fertilizer on yield of conventional tillage and no-tillage corn at eight locations in Kentucky. (From Frye *et al.*, 1981.)

		FERTILIZER NITROGEN APPLIED			
		85 kg/ha		170 kg/ha	
SOIL	TIME OF APPLICATION[3]	CONVENTIONAL	NO-TILL	CONVENTIONAL	NO-TILL
		Yield of Grain (kg/ha)			
Allegheny[1]	p	–	8,906	–	10,412
	d	–	9,533	–	10,286
Baxter[1]	p	9,283	11,039	9,283	11,039
	d	9,910	10,788	9,533	11,540
Cavode[2]	p	8,154	8,655	8,279	9,784
	d	7,276	8,781	7,903	9,283
Hampshire[2]	p	–	–	–	6,523
	d	–	–	–	8,216
Monongahela[2]	p	9,032	6,969	9,032	10,600
	d	8,467	7,965	8,781	8,216
Lowell[1]	p	–	8,530	–	12,356
	d	–	10,192	–	11,346
Tilsit[2]	p	–	6,648	–	8,298
	d	–	7,495	–	7,890
Tilsit[2]	p	–	6,742	–	8,060
	d	–	8,060	–	9,791

[1] Classed as well-drained soils but tend to be wet in spring due to fine texture, especially under no-tillage.

[2] Somewhat poorly drained soils.

[3] p indicates nitrogen fertilizer applied at planting; d indicates nitrogen fertilizer applied four to six weeks after planting.

negative responses than positive ones to delayed application of 170 kg/ha nitrogen for both conventional tillage and no-tillage. Delayed application of nitrogen fertilizer increased yields more frequently on the somewhat poorly drained soils than on the well-drained soils. This is as would be expected in view of the potential for greater losses of nitrogen through denitrification from these soils.

In Kentucky, it is recommended that nitrogen fertilizer for corn be decreased by 40 kg/ha if as much as two-thirds of the nitrogen is applied four to six weeks after planting no-tillage corn on moderately well-drained soils or conventional tillage corn on moderately well- and poorly-drained soils (no-tillage on poorly-drained soils is not recommended in Kentucky). The nitrogen fertilizer saved by the improved efficiency represents about 2,300 MJ/ha of energy. Usually an extra trip over the field is required to make the delayed application, which offsets an estimated 80 MJ/ha of the energy saved. Therefore, based on the fertilizer recommendations for corn in Kentucky, delayed application of nitrogen fertilizer on soils where it is needed can conserve about 2,220 MJ/ha of energy due to more efficient use of nitrogen. In terms of DFE, this is about 54 liters/ha. Thus, delayed application of nitrogen fertilizer on wet soils can be a significant energy-conserving practice in most cases with no-tillage and in some cases with conventional tillage.

Effect of Fertilizer Placement on Efficiency

Several studies in recent years have indicated that both nitrogen and phosphorus fertilizers are used more efficiently by the crops when applied in a band below the soil surface, compared to broadcast application on the soil surface (Murphy, 1981). Subsurface placement of fertilizers requires substantially more energy for the application operation than surface application; however, the significant yield increases that have been reported may make the additional energy use worthwhile. More research is needed on this practice.

Nitrogen From Legume Cover Crops

Legume cover crops can be used to provide at least part of the nitrogen requirement for no-tillage corn, resulting in a substantial reduction in the need for nitrogen fertilizer. Results of corn grain yields in Kentucky with three winter annual cover crops and rye compared to corn stalk residue are shown in Table 6-11. Corn yields at all nitrogen fertilizer rates following a hairy vetch cover crop were considerably greater than with rye cover. We estimated that hairy vetch provided about 90 kg/ha nitrogen for the corn that year. Crimson clover and big flower vetch provided lower amounts. Assuming 90 kg/ha as the potential to provide nitrogen by winter legume cover crops, the amount of energy which could be saved due to lower nitrogen fertilizer rates would be about 5,200 MJ/ha, equivalent to 130 liters/ha diesel fuel.

The Office of Planning and Evaluation of the U.S. Department of Agriculture (1975) estimated that 65 percent of the production of seven important row crops, including corn, will be grown by no-tillage by the year 2000. Assuming that a total of 35 million ha of corn will be grown in the U.S. by the year 2000 and winter legume cover crops will be used on one-half the no-tillage corn, the amount of nitrogen saved would be approximately 1 billion kg/yr. This would represent approximately 58 billion MJ of energy saved annually. Some energy would be required to plant the legumes, and the legume seeds would represent a small input of energy. Subtracting estimated values of 300 and 200 MJ/ha, respectively, for these energy inputs, the amount of energy that could be saved annually by using legume cover crops for no-tillage corn would be approximately 52 billion MJ. This is equivalent to about 1.3 billion liters of diesel fuel. Such a saving of energy could be important to the individual farmer and make a

Table 6-11. Effect of winter legume cover crops on yield of no-tillage corn at Lexington, Kentucky, 1977-81; values are averages of five years.

	N APPLIED (kg/ha)		
WINTER COVER	0	50	100
	Yield of Grain (kg/ha)		
Corn stalk residue	3,780	5,220	6,820
Rye	4,020	5,710	7,590
Crimson clover	4,440	5,600	7,360
Big flower vetch	4,220	6,630	6,570
Hairy vetch	6,420	7,450	9,020

contribution to the conservation of energy used in agriculture, although, in terms of the total U.S. energy demand, it is very small.

Results obtained by others in areas with milder winters than Kentucky indicated larger amounts of nitrogen provided by legume cover crops. In Delaware, researchers estimated that winter cover crops consisting of mixtures of hairy vetch and small grains or crimson clover and small grains resulted in corn yields comparable to those obtained with 112 kg/ha nitrogen applied as nitrogen fertilizer without legumes (Mitchell and Teel, 1977).

It should be pointed out that the energy conservation values discussed here are small by comparison when all of the ways in which legume crops can be used to supply nitrogen are considered. If one took into account the potential for using legume cover crops for conventionally tilled corn and other crops, legumes in rotation with nonlegumes, and legumes in association with pasture and hay grasses, the energy savings would be much greater.

ENERGY EFFICIENCY IN FORAGE PRODUCTION

To obtain optimum yields of forage grasses, a rather large amount of nitrogen is required. As with corn, the nitrogen could represent the largest energy input into their production if provided by nitrogen fertilizer. The quantity of production is generally proportional to the amount of nitrogen available to the plant up to the optimum level. Legume plants can be used to provide the nitrogen needed by forage grasses and, at the same time, improve the quality of the forage compared to grasses alone.

Grassland renovation or pasture improvement is achieved primarily through the interseeding of legumes into grass sods. The conventional method of grassland renovation is usually disking three times, planting legumes with a grain drill, followed by one trip over the field with a cultipacker to firm the seedbed. No-tillage renovation involves opening slits 1–2 cm wide and about 2 cm deep in rows about 20 cm apart. Legume seeds are planted in the slits. Less than 10 percent of the soil surface is disturbed in the no-tillage method of renovation.

There are two important aspects to energy conservation potential through grassland renovation. One is in providing nitrogen for high yields and improved quality forage by establishing or re-establishing legumes in the sward instead of using nitrogen fertilizer. The other is by interseeding the legumes into the grass using the no-tillage method of renovation. Research results have shown that the two combined make no-tillage renovation of grassland a significant energy-conserving practice in grassland farming.

Experiments in Kentucky in which red clover was interseeded into plots of fescue grass (*Festuca arundinacea* Shreb.) showed that forage yields of the mixture without nitrogen fertilizer were as great as fescue alone with 112 kg/ha nitrogen fertilizer during the establishment year (Taylor, 1975). During the second year, the red clover-fescue mixture without nitrogen fertilizer yielded as much as fescue alone with 224 kg/ha nitrogen fertilizer. Based on these results, Table 6-12 shows a comparison of estimated energy requirements for

Table 6-12. Comparison of energy requirements for improvement of fescue grass stands with nitrogen fertilizer or with red clover established conventionally or by no-tillage. (From Taylor, 1975.)

	METHOD OF IMPROVEMENT		
		RED CLOVER[3]	
MANAGEMENT INPUT OR OPERATION	N FERTILIZER	CONVENTIONAL[1]	NO-TILL[2]
		(liters/ha/yr)	
Disking (three times)	–	8.9	–
Seeding	–	2.3	4.8
Seed (6.8 kg/ha)	–	1.4	1.4
N fertilizer (170 kg/ha N)	240	–	–
Machinery (manufacture and repair)	–	1.6	0.3
Herbicides (manufacture)	–	–	1.0[4]
Apply fertilizer (granular)	1	1.0	1.0
Totals	241	15.2	7.5

[1] Conventional renovation consisted of disking three times, seeding red clover with a grain drill, followed by one trip over with a cultipacker.

[2] No-tillage renovation consisted of seeding red clover with a grassland renovation seeder manufactured for use with no-tillage.

[3] Values averaged over two years, the expected lifespan of a stand of red clover.

[4] Close grazing or mowing may be substituted for herbicides at establishment.

improvement of fescue grass stands by applying nitrogen fertilizer or by establishing red clover conventionally or using a no-tillage method. Establishing legumes in grass stands using a no-tillage grassland renovation seeder conserves a substantial amount of energy over the conventional method. If the energy represented by phosphorus and potassium fertilizers and lime is ignored, the energy requirement for conventional interseeding of the legumes is about twice that of no-tillage. However, the energy required by either interseeding method is very small in comparison to forage grass improvement by applying nitrogen fertilizer to an existing stand. These results emphasize the role which legumes can play in nitrogen efficiency in cropping systems.

REFERENCES

Achorn, F. P. 1981. Energy equation. Part 1. Producing, transporting, storing and applying N – what does it take? *Solutions* (July/August), pp. 16–28. Nat. Fert. Solutions Assoc., Peoria, Illinois.

American Chemical Society. 1974. Agriculture depends heavily on energy. *Chem. Eng. News.* **52**:(*10*):23–24.

Collins, N. E., T. H. Williams, and L. J. Kemble. 1980. Measured machine energy requirements for grain production systems. In *Agricultural Energy, Vol. 2. Biomass Energy-Crop Production.* Am. Soc. Agric. Engr., St. Joseph, Michigan.

d'Arge, Ralph C. 1981. The energy squeeze and agricultural growth. Walter E. Jeske, Editor. *Economics, Ethics, Ecology: Roots of Productive Conservation.* Soil Conservation Society of America, Ankeny, Iowa.

Energy Information Center. 1982. Personal communication. Office of Energy Information Services. U.S. Department of Energy. Washington, D.C.

Frye, W. W., R. L. Blevins, L. W. Murdock, and K. L. Wells. 1981. Energy conservation in no-tillage production of corn. In *Crop Production with Conservation in the 80's.* (Presented in Chicago.) Am. Soc. Agric. Engr., St. Joseph, Michigan ASAE Pub. 7-*81*:255–262.

Frye, W. W. and S. H. Phillips. 1980. How to grow crops with less energy. In *Cutting Energy Costs. The 1980 Yearbook of Agriculture.* U.S. Department of Agriculture, Washington D.C.

Griffith, D. R. and S. D. Parsons. 1980. Energy requirements for various tillage-planting systems. (Tillage) ID-141. Cooperative Extension Service, Purdue University, W. Lafayette, Indiana.

Heichel, G. H. 1976. Agricultural production and energy resources. *American Scient.* **64**:64–72.

Mitchell, W. H. and M. R. Teel. 1977. Winter-annual cover crops for no-tillage corn production. *Agron. J.* **69**:569–573.

Moschler, W. W. and D. C. Martens. 1975. Nitrogen, phosphorus, and potassium requirements in no-tillage and conventionally tilled corn. *Soil Sci. Soc. Am. Proc.* **39**:886–891.

Murphy, L. 1981. Agriculture is net producer of energy. *Solutions* (July/August), pp. 48–56. Nat. Fert. Solutions Assoc., Peoria, Illinois.

Na Nagara, T. and R. E. Phillips. 1971. Unpublished data. Department of Agronomy, Kentucky Agric. Exp. Sta., Lexington, Kentucky.

National Research Council. 1977. World food and nutrition study. National Academy of Sciences, Washington, D.C.

Nelson, L. W. 1975. Fertilizers for all-out food production. W. P. Martin, Editor. *All-Out Food Production: Strategy and Resource Implications.* ASA Spec. Pub. No. 23. American Society of Agronomy, Madison, Wisconsin.

Parsons, Samuel D. 1980. Estimating fuel requirements for field operations. AE-110. Cooperative Extension Service, Purdue University, W. Lafayette, Indiana.

Phillips, R. E., R. L. Blevins, G. W. Thomas, W. W. Frye, and S. H. Phillips. 1980. No-tillage agriculture. *Science* **208**:1108–1113.

Shelton, D. P. 1980. Factors influencing specific fuel use in Nebraska. In *Agricultural Energy, Vol. 2. Biomass Energy-Crop Production.* Am. Soc. Agric. Engr., St. Joseph, Michigan.

Soil Conservation Society of America. 1978. Energy and conservation of renewable resources. Soil Conservation Society of America, Ankeny, Iowa.

Stanford, G., V. A. Bandel, J. J. Meisinger, and J. O. Legg. 1979. N behavior under no-till and conventional corn culture. II. Grain and forage yields in relation to amounts of N applied and total N uptake. *Agron. Abs.* (1979), p. 183.

Steinhart, J. S. and C. E. Steinhart. 1974. Energy use in the U.S. food system. *Science* **184**:307–316.

Stout, B. A. 1979. *Energy for World Agriculture.* Food and Agriculture Organization of the U.N., Rome, Italy.

Taylor, T. H. 1975. Establishing legumes in tall fescue sod. Paper presented at Southern Beef Cattle Conf., Greenville, South Carolina.

USDA Economic Research Service/Federal Energy Administration. 1977. Energy and U.S. agriculture 1974 data base. FEA/0-77/140. U.S. Government Printing Office, Washington, D.C.

U.S. Department of Agriculture, Office of Planning and Evaluation. 1975. *Minimum Tillage: A Preliminary Technology Assessment.* U.S. Government Printing Office, Pub. No. 57-398, Washington, D.C.

Wittmuss, H., L. Olson, and D. Lane. 1975. Energy requirements for conventional versus minimum tillage. *J. Soil and Water Cons.* **30**:72–75.

Wittmuss, H. D. and A. Yazar. 1981. Moisture storage, water use and corn yields for seven tillage systems under water stress. In *Crop Production with Conservation in the 80's.* (Presented in Chicago.) Am. Soc. Agric. Engr., St. Joseph, Michigan, ASAE Pub. **7**–*81*:66–75.

Wittwer, S. H. 1975. Food production: Technology and resource base. *Science* **188**:579–584.

Wittwer, S. H. 1981. Resource conservation-agricultural productivity. In *Crop Production with Conservation in the 80's.* (Presented in Chicago). Am. Soc. Agric. Engr., St. Joseph, Michigan, ASAE Pub. 7-*81*:1–15.

7
Response of Weeds and Herbicides Under No-Tillage Conditions

William W. Witt
Associate Professor of Agronomy
University of Kentucky

SOME FACTORS AFFECTING WEED CONTROL IN NO-TILLAGE

The acceptance of no-tillage production of agricultural crops has been dependent on the development of herbicides for providing suitable weed control. Although cultivation is not always impossible in no-tillage production, it is rarely a viable weed control alternative. This places an increased demand on the herbicide to provide consistent, season-long control.

The presence of weeds in cultivated crops has been shown to reduce crop quantity and quality and cause competition for available soil water and nutrients and interference with harvesting efficiency. These deleterious effects of uncontrolled weeds can occur regardless of the tillage system in which the crop is grown. However, the presence of weeds in these two systems can result, and often does, in different effects on crop production over a multi-year period. Likewise, when herbicides are introduced into these two environments, the resultant weed control and duration of persistence can differ significantly. Also, the unique ability of weed species to reproduce and maintain the species over time is altered in no-till and conventional till production systems.

A primary reason for soil tillage is to control weeds. Prior to the discovery of herbicides, weed control was accomplished by preplanting tillage operations, multiple between row cultivations, and hand weeding. Cultivations were usually timed to remove weeds when they were small. Obviously, annual weeds could be removed with fewer cultivations than could perennial species. Pavlychenko (1944) pointed out that "any measure devised for weed control, to be practical, must not harm the crops, should effectively destroy the weeds, and at the same time conserve soil moisture, control soil drifting, and must be economical to the farmer."

He found that planting grain crops into previous years' stubble without any pre-seeding cultivation allowed shallow germinating weeds to begin growth where they could effectively be removed by mechanical cultivation. Wheat (*Triticum aestivum* L.), rye (*Secale cereale* L.), oats (*Avena sativa* L.), and barley (*Hordeum vulgare* L.), which were seeded deeper, would emerge from the soil after the weeds and could effectively outgrow, or "out-compete" the weeds during the growing season. This type of system worked well for shallow germinating weeds, but not for deeper germinating weeds nor for deep-rooted perennial species. Since this research was conducted, this one overriding theme has emerged in practically all reduced tillage systems – perennial species are much more difficult to control, even with the multitude of herbicides available today.

Multiple tillages for weed control have resulted in severe erosion (Hanson, 1944) on many soils, and attempts to control weeds with herbicides has led to much research which has documented that many weed species can be controlled with herbicides, either in crops during the growing season, or on fallow land (Wiese and Army, 1960; Wiese, 1956).

Herbicides have been shown to provide acceptable weed control, without, or with reduced tillage, in corn (*Zea Mays* L.) (McClure *et al.,* 1968, Shear, 1965, Klingman and Spain, 1965, McKibben, 1979, Moomaw and Martin, 1976, Fenster and Robinson, 1970, Griffith, 1970, Williams and Ross, 1970, Herron and Phillips, 1969); soybean (*Glycine max* [L.] Merrill) (Herron *et al.*, 1973, Henard *et al.,* 1969, McKibben, 1970, Ross and Williams, 1969, Worsham, 1970); alfalfa (*Medicago sativa* L.) (Faix *et al.,* 1977, Herron *et al.*, 1974, Faix *et al.,* 1979; red clover (*Trifolium pratense* L.) (Peters and Lowance, 1977), pearl millett (*Pennisetum americanum* [L.] Leeke) (Peters and Lowance, 1970), sunflower (*Helianthus annuus* L.) (Kosovac, 1968), and tobacco (*Nicotiana tabacum* L.) (Chappel and Link, 1977).

Although herbicides can be used effectively for weed control in no-tillage crop production, other factors associated with weeds and herbicides merit consideration. These are (1) what effect the lack of tillage has on weed seed distribution on the soil surface and within the soil profile in relation to weed seed viability and possible species shifts over time and (2) what effect the accumulated mulch on the soil surface has on herbicide activity and persistence.

Many types of reduced tillage systems have been developed for production of agricultural commodities. This chapter will deal primarily with those practices in which only minimal tillage has occurred. For the most part this will be called no-tillage, in which the only tillage operation is by the planting equipment. Where possible, this will be compared to a conventional system of moldboard plowing and subsequent tillage for seedbed preparation, or with various types of minimum tillage.

The consequences of tillage, or a lack thereof, and the response of weeds to this influence was reviewed by Cussans (1966). He noted that tillage (plowing) had three basic effects. These were (1) the burial of surface vegetation and weed seeds, (2) inversion of the soil so that weed seeds previously buried are brought to the surface and (3) disturbance of the soil, by either a primary tillage operation, or subsequent secondary tillage (between row cultivation), which can shatter rhizome and root systems and be brought to the surface, and by which weed seeds can be brought to the surface to an environment suitable for germination.

Under no-tillage conditions, the inversion and disturbance phases will be kept to a minimum. Obviously, slight disturbance will occur by the planting equipment.

Plant material, either alive or dead, on the soil surface can influence germination of both weed and crop seed. Elliot (1974) points out that the use of herbicides, which made no-till seeding possible, is being altered in the face of a changing weed flora brought about by a lack of tillage. He categorized the response of certain weed types in direct seeded cereal grains as follows: (1) Annual broadleaved weeds – the less soil disturbance, the less are the numbers of the weeds that emerge, but enough do emerge to require selective herbicide applications. Chickweed (*Stellaria media* [L.] Cyrillo), mayweed (*Matricaeia* spp) and parsley piert (*Aphanes arvensis*) have been observed to increase in direct seedings. (2) Perennial grasses – quackgrass (*Agropyron rapens* [L.] Beauv.) increased in no-tillage situations. (3) Annual grasses – direct drilling of cereal grains appeared to allow for easier control of wild oat (*Avena fatua* L.) because of the wild oat seed being left exposed on the soil surface.

The effect of four years of direct drilling on types of weeds present in cereal grains was reported by Bachthaler (1974). A large increase in quackgrass infestation was found under direct drill planting (Table 7-1), while the annual grasses increased primarily with direct drilling of oat. It is interesting to note that annual broadleaf weeds increased in plots that were tilled in the fall wheat and barley. The number of weed species present was approximately

Table 7-1. Structure of the weed flora in the fourth experimental year. Average results of three mineral soils; locations: Duerrnhof, Herbstadt and Mantlach. (From Bachthaler, 1974.)

	WINTER WHEAT AND SPRING WHEAT		SPRING BARLEY		OATS	
	SOWN AFTER PLOUGH	DIRECT DRILL	SOWN AFTER PLOUGH	DIRECT DRILL	SOWN AFTER PLOUGH	DIRECT DRILL
Broad-leaved annual weeds	2	+	1	+	+	+
Broad-leaved perennial weeds	+	+	+	+	1	1
Annual grass weeds	+	+	+	1	1	2–3
Perennial grass weeds (*Agropyron repens*)	1	3	+	2–3	1	2
Number of weed species	16	16	17	13	16	13

Percent weed cover
+ = below 1%
1 = 1–5%
2 = 5%–25%
3 = 25%–50%
4 = 50%+

the same under both tillages. However, there was a tendency for fewer species present under direct drilling.

Pollard and Cussans (1976) reported results of an in-depth study of tillage effects on specific weed species over a multi-year period in cereals (Table 7-2). Species occurrence and density varied considerably at each location for each of the years.

The large number of species reported demonstrates the complex plant ecosystem found in field situations. Determinations of the modifying properties of tillage and/or herbicides on such an ecosystem are likewise complex. Any attempt at determining shifts, or changes, in weed species spectrum must be measured in detail each year and the experiments conducted over multiple year periods.

At the Begbroke II location, at which data was collected and reported, the number of weeds present varied from year to year. At all locations, the number of species increased each succeeding year. The number of dicotyledonous weed seedlings demonstrated effect of tillage, or lack of, on weed density (Table 7-3).

An important result of their research was the reduced emergence of broadleaf seedlings under no-tillage conditions compared with plowing. They found no continuous long-term trends caused by tillage in populations of broadleaf weeds. The ratios between seedling numbers on plowed plots and numbers on no-tilled plots remained relatively steady and no tendency for differences to increase with time might be expected. They presumed that long-term trends of this type are diluted by large reserves of dormant seeds in the soil. They categorized individual weed species response into the following five categories: (1) species increased by tillage, (2) species reduced by plowing, (3) species increased by timed cultivation, (4) species showing an inconsistent response and (5) species showing no response.

Results similar to those reported earlier by Bachthaler (1974) were found with more broadleaf weed seedlings present with increased levels of tillage. Annual grasses such as *Avena fatua* and *Alopecurus myosuroides* were favored by reduced cultivation. Generally, perennial species were more prevalent under no-tillage.

Several researchers have found that annual weeds are easier to control in no-tillage plots compared to perennial species. Sarpe (1974) reported that corn could be successfully grown under no-tillage and minimal-tillage conditions except where perennial weeds such as

Table 7-2. The occurrence and abundance of seedling weed species at four locations in differing years. (From Pollard and Cussans, 1976.)

	BEGBROKE I			BEGBROKE II					COMPTON		BUCKLAND
	1968	1969	1970	1969	1970	1971	1972	1973	1970	1973	1973
Alopecurus myosuroides											4
Anagalis arvensis							3	2		1	1
Aphanes arvensis				1		1	1	1			
Avena fatua			1					1			
Capsella bursa-pastoris	3		2		1		1	1			1
Cerastium spp		2				1	1	1	1	1	1
Chenopodium album					2	1	1	1		1	1
Chrysanthemum segetum			2								
Cirsium spp							1	1	1	1	2
Epilobium spp								1			1
Euphorbia spp								1			
Fumaria officinalis	1	1	1	1			1	1	1		
Galium aparine						1		1		1	
Hypochaeris radicata								1		1	1
Lamium amplexicaule										1	1
Lapsana communis								1			
"Mayweeds"			3		1	1	1	2			1
Papaver rhoeas								1	1		1
Poa annua								3		2	
Polygonum aviculare	3	4	5	3	4	4	4	4	2	2	1
Polygonum convolvulus			2		3	2		2	3	2	1
Polygonum lapathifolium								2			
Polygonum persicaria		4	2	2	3	2		1		1	
Ranunculus spp			2			1	1	1			
Raphanus raphanistrum + Sinapis arvensis	5	5	5	3	3	2	3	2			1

Table 7-2. The occurrence and abundance of seedling weed species at four locations in differing years. (From Pollard and Cussans, 1976.) (Continued)

	BEGBROKE I			BEGBROKE II					COMPTON		BUCKLAND
	1968	1969	1970	1969	1970	1971	1972	1973	1970	1973	1973
Rumex spp			3				1				
Sambucus nigra							1	1	3	2	
Senecio vulgaris	3	3	4	1	2	1	2	1			1
Sisymbrium officinale							1				
Solanum nigrum							1				
Sonchus spp		3	3			1	2	2		1	3
Spergula arvensis		3	3		1		1				
Stellaria media		2	1	1	2	1	2	2	3	2	3
Taraxacum officinale		3					1	1			1
Trifolium repens							2	2	1	3	1
Urtica urens								1			
Veronica persica							1		3	2	1
Viola arvensis	4	4	5	1	2		1	1	2	1	
Total seedling weeds	5	6	6	4	5	4	5	4	4	4	5
Number species recorded	6	11	14	9	11	13	23	29	11	17	21

1 = <1 plant/m^2; 2 = 1.01–3.16; 3 = 3.17–10.0; 4 = 10.1–31.6; 5 = 31.7–100; 6 = >100 plants/m^2.

Table 7-3. Total dicotyledonous seedlings under different tillage systems. NT = no tillage, ST = shallow tine cultivation, DT = deep tine cultivation, P = autumn or winter plowing. (From Pollard and Cussans, 1976.)

	YEAR	NT	ST	DT	P
			(plant/m^2)		
Begbroke I	1968	12.1	48.1	62.7	–
	69	19.6	107.5	204.6	–
	70	31.0	139.7	215.5	–
Begbroke II	69	2.9	4.7	6.9	18.3
	70	9.6	28.3	37.2	35.5
	71	7.9	11.5	16.0	28.0
	72	8.1	9.3	19.1	32.8
	73	6.5	26.1	17.8	23.2
Compton	70	21.7	22.2	25.2	24.0
	73	7.0	8.3	16.5	17.1
Buckland	73	15.3	5.4	2.2	1.8

johnsongrass (*Sorghum halepense* [L.] Pers.) were present. Scharbau (1968) reported excellent control of annual weeds with paraquat in much of Europe and concluded that this treatment could replace the weed control function of tillage in many instances.

Weed emergence from the soil, in either disturbed or undisturbed conditions, can be influenced by the predominant species in any one particular location. The importance of the very large weed seed reservoir found in most agricultural soils cannot be overlooked when discussing emergence of any weed species. Roberts and Dawkins (1967) and Roberts and Feast (1973) determined the rate of decrease of seed in disturbed and undisturbed soils. The numbers of viable seeds decreased from year to year in an exponential manner (Figure 7-1). The undisturbed bare surface, the undisturbed mulch (farmyard manure) and undisturbed with grass sod resulted in decreases of 34, 31 and 32 percent per year, respectively. On plots tilled twice a year the loss in viable seed was 42 percent per year and on those plots tilled seven times per year resulted in a loss of 56 percent per year. This value was significantly greater than for any other treatment. Results reported by Roberts and Dawkins (1967) were essentially the same for two treatments that were the same in both experiments. These results are from locations that did not receive a

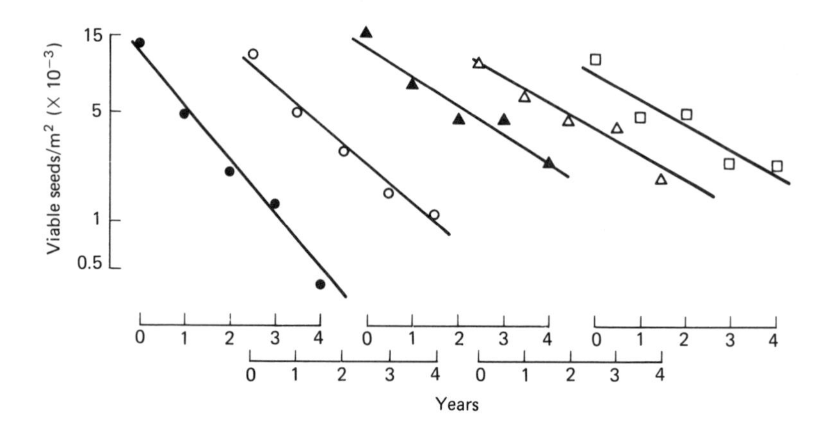

Figure 7-1. Numbers of viable seeds all species together, in the top 23 cm of soil. Tilled seven times a year; 0, tilled twice a year; □, undisturbed, surface bare; ■, undisturbed, mulched; △, undisturbed, grass sod (from Roberts and Feast, 1973).

fresh influx of weed seeds each year. Obviously, such results would differ considerably when more weed seeds are available to reinfest soils under actual cropping situations.

Roberts and Feast (1972) reported the results of seedlings emerged as a percent of weed seeds added to soil at different depths. Undisturbed samples were left in place until removed for analysis. The disturbed samples were removed and incorporated to the various depths four times per year. Initial seedling emergence during the first of all species together amounted to 22, 11 and 3 percent of the seeds added for the 2.5-, 7.5- and 15-cm depths, respectively, for the cultivated soil; for the undisturbed soil the comparable values were 16, 6 and 2 percent, respectively.

Fewer seedlings emerged in the undisturbed soil compared to the cultivated soil (Table 7-4). The first year, greatest emergence occurred from the shallowest depth for both systems. In the second year, emergence was essentially the same from each depth in the cultivated soil but in undisturbed soil, greater emergence occurred from seeds in the shallower depths. In the final three years of the study, more emergence occurred with increasing depth of incorporation. There was little difference in the final two years in the undisturbed soil, regardless of depth of incorporation.

They also reported that, although the number of weeds emerging was less in undisturbed soil, the number of viable seeds remaining in

Table 7-4. Yearly emergence of seedlings of annual weed from seeds mixed with cultivated and undisturbed soil from differing soil depths. (From Roberts and Feast, 1972.)

	SEEDLINGS AS % OF SEEDS ADDED					
	CULTIVATED			UNDISTURBED		
YEAR	2.5 cm	7.5 cm	15 cm	2.5 cm	7.5 cm	15 cm
1	52.6	36.7	25.1	36.8	19.5	10.2
2	16.3	17.6	13.7	12.6	9.0	5.7
3	3.9	6.1	8.4	4.8	3.6	2.7
4	1.7	3.1	4.6	2.2	2.4	1.3
5	0.9	1.5	2.6	1.6	1.7	1.3

the soil was much greater. This phenomenon is often not realized in many areas where crops may be grown under no-tillage conditions for one or two years, and then tilled. Viable seeds will be brought to, or near to, the surface, where a favorable germination environment exists. Any reduction in weed seedling emergence in a no-till system will be an advantage only as long as the soil is left undisturbed.

Another complicating factor that should be considered is that seed of a particular weed species usually does not behave like another in a similar environment. Data from Roberts and Feast (1972) clearly indicate the variability by weed species in emergence from different depths of incorporation (Table 7-5).

EFFECT OF SURFACE PLANT RESIDUE ON HERBICIDE PERSISTENCE

Plant residue on the soil surface has been a subject of interest from the standpoint of its effect on herbicides and subsequent weed control. The amount of surface residue will vary greatly depending on the type of plant material and whether it is alive or was remaining from a previous crop. As discussed earlier in this chapter, crops have been grown successfully with no-tillage practices under a wide variety of plant residue conditions. Witt (1981) discussed weed control in relation to surface residue and problems that can be encountered as a result of the residue interfering with adequate coverage by the herbicidal spray mixture. Erbach and Lovely (1975) found no

Table 7-5. Fate of seeds of annual weeds when mixed with cultivated and undisturbed soil from differing soil depths. (From Roberts and Feast, 1972.)

	CULTIVATED						UNDISTURBED					
	% EMERGED DURING 5 YEARS			% RECOVERED AFTER 5 YEARS			% EMERGED DURING 5 YEARS			% RECOVERED AFTER 5 YEARS		
WEED SPECIES	2.5 cm	7.5 cm	15 cm	2.5 cm	7.5 cm	15 cm	2.5 cm	7.5 cm	15 cm	2.5 cm	7.5 cm	15 cm
Capsella bursa-pastoris	77	48	51	2	5	6	56	29	17	3	12	26
Chenopodium album	83	73	48	5	9	16	42	31	11	24	30	66
Euphorbia helioscopia	88	82	76	1	3	2	84	57	30	7	19	31
Fumaria officinalis	85	74	70	4	6	13	60	50	31	12	31	43
Matricaria matricarioides	49	32	23	3	8	14	34	13	9	6	14	19
Medicago lupulina	82	50	39	6	12	17	57	43	20	12	21	29
Papaver rhoeas	64	45	37	4	2	9	20	18	7	8	14	17
Poa annua	71	68	63	1	5	22	52	41	28	4	11	29
Polygonum aviculare	50	46	34	0	2	4	34	20	9	5	16	42
Polygonum convolvulus	73	67	50	0	2	3	66	51	22	4	7	24
Senecio vulgaris	100	67	48	0	0	0	81	47	26	0	2	5
Spergula arvensis	68	58	38	1	1	1	52	36	15	3	4	17
Stellaria media	65	69	53	0	2	2	64	30	19	1	6	15
Thlaspi arvense	81	79	70	0	2	3	65	40	24	12	22	60
Tripleurospermum maritimum spp. inodorum	52	36	32	6	7	11	34	20	8	8	20	27
Urtica urens	67	62	52	2	3	10	54	26	14	7	27	39
Veronica hederifolia	87	82	96	0	0	0	78	23	29	3	31	51
Veronica persica	89	90	74	1	1	2	79	29	26	3	11	29
Vicia hirsuta	91	94	76	3	2	11	96	85	57	4	9	17
Viola arvensis	88	80	61	6	7	8	52	36	17	12	23	44

differences in control of foxtail millet (*Seteria italica* [L.] Beauv.) or velvetleaf (*Abutilon theophrasti* Medic.) with up to 4,000 kg/ha of surface corn residue, provided that herbicides were applied at rates recommended for that particular soil. However, at less than recommended rates, surface residue did cause a reduction in weed control.

Herbicide persistence in no-tillage soil systems, irrespective of type of no-tillage, is dependent on the chemical properties of the herbicide, rate of application, soil pH, soil organic matter content, amount of surface plant residue, temperature, rainfall and microbial decomposition. With many management systems for no-tillage crop production, the way in which the surface plant residue is managed also varies. Continuous no-tillage corn production is known to reduce soil pH compared to conventionally tilled corn (Blevins *et al.*, 1977). The effect of soil pH on triazine herbicides has been investigated under several production systems. Hiltbold and Buchanan (1977) reported increased atrazine [2-chloro-4-(ethylamino)-6-(isopropylamino)-*s*-triazine] persistence under field conditions as pH increased in conventionally tilled soils. Lowder and Weber (1979) reported increased atrazine persistence in no-tillage plots which were limed compared to those unlimed. Schnappinger *et al.* (1977) reported decreased weed control from atrazine and simazine [2-chloro-4,6-bis(ethylamino)-*s*-triazine] in no-tillage corn where surface nitrogen solutions had produced increased acidity in the top 1.2 cm of soil. Similar results were reported for simazine under no-tillage conditions by Kells *et al.* (1979). Simazine provided less weed control under acidic surface conditions and the effect was more pronounced under no-tillage than conventional tillage. Slack *et al.* (1978) reported less simazine persistence under no-tillage corn than under conventional tillage. Oat bioassay of soils showed lowest levels of simazine in soils with a pH of 5.4 or lower (Figure 7-2). Results were similar for data collected in 1975 and 1976. Total precipitation during the growing season was 47.6 cm in 1975 and 50.6 cm in 1976. They concluded that the more rapid decrease in phytotoxic effect of simazine under no-tillage as compared with conventional tillage could be due to at least three factors: (1) higher organic matter content of the no-tillage soil sample zone (0–8 cm) studied; (2) adsorption effect of the organic residue on the surface at time of application; and (3) the higher moisture content in the surface soil of no-tillage plots. The no-tillage plots had a range of soil organic matter content of 4.3–5.1 percent compared to a range of 2.7–3.1 percent for conventionally tilled plots. The significance of reduced persistence of simazine, and other chloro-triazines should not be overlooked by corn producers utilizing no-tillage planting. Timely lime application would probably be needed, or increased rates of herbicides to achieve weed control.

In a detailed study of three chloro-*s*-triazines, Slack (1973) found that at any given time after application, less herbicide was found under no-tillage than under conventional tillage. He reported cyanazine [2-[[4-chloro-6(ethylamino)-*s*-triazine-2-yl]amino]-2-methylpropionitrile] to be the least persistent with little herbicide remaining for over 60 days. Atrazine persisted up to 120 days, depending on the rate applied and simazine persisted over 120 days at all rates except the 1.7 kg/ha rate (Table 7-6).

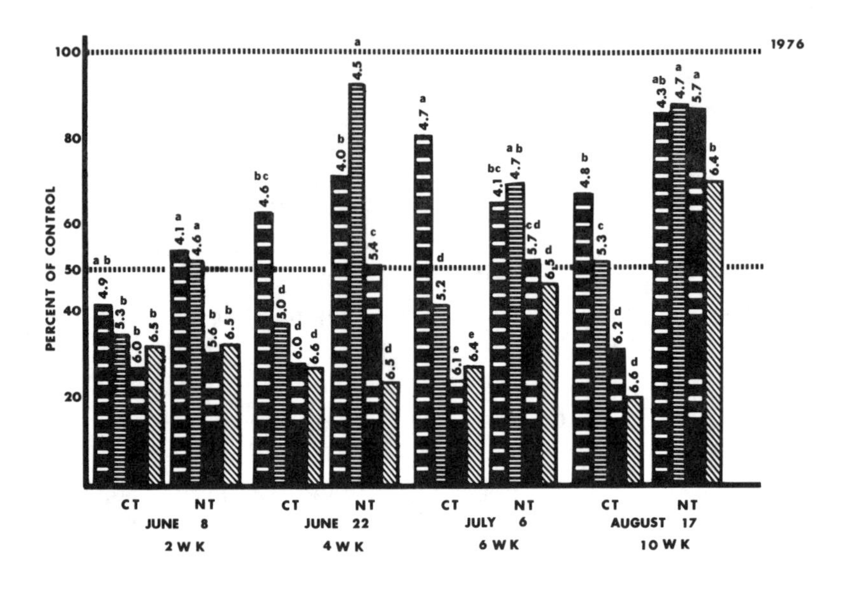

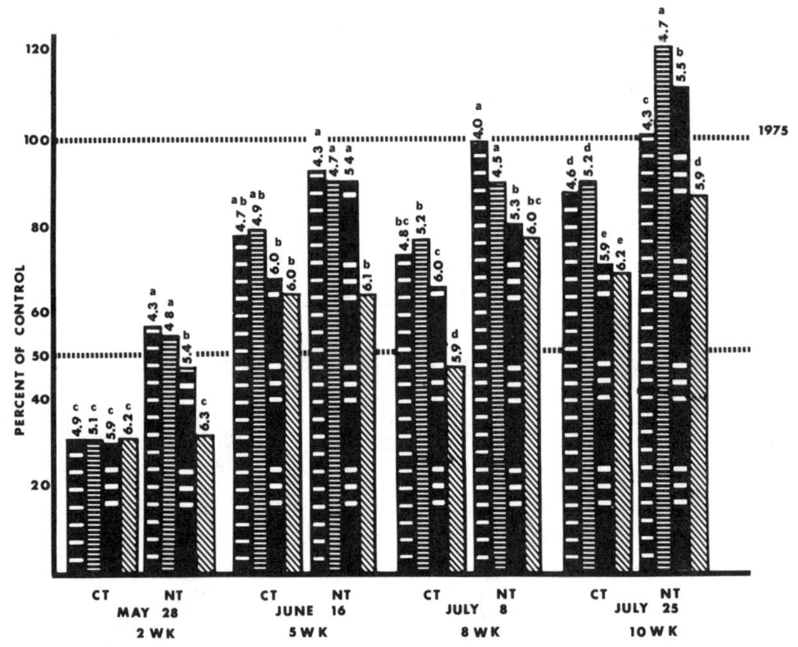

Figure 7-2. Biomass data of dry weight yield of oats expressed as a percent of control treatments with different pH levels under no-tillage (NT) and conventionally-tilled (CT) management sampled at different times during the 1975 and 1976 growing seasons; means within sampling dates with similar letters are not significantly different at the 5 percent level using Duncan's multiple range test (from Slack *et al.*, 1978).

Table 7-6. Response of oat and barley as a cover crop 124 days after chloro-s-triazine application. (From Slack, 1973.)

		CROP INJURY (%)			
		TILLED		NO-TILLED	
HERBICIDE	RATE (kg/ha)	OAT	BARLEY	OAT	BARLEY
Atrazine	1.7	0	0	0	0
	3.3	3	3	0	0
	6.6	23	18	0	0
	13.2	95	93	60	38
Simazine	1.7	5	5	0	0
	3.3	28	23	5	0
	6.6	78	68	60	43
	13.2	95	93	100	100
Cyanazine	1.7	0	0	0	0
	3.3	0	0	0	0
	6.6	0	0	0	0
	13.2	0	0	0	0

Kells *et al.* (1980), utilizing ^{14}C-atrazine, reported less atrazine persistence under no-tillage compared with conventionally tilled soils and related persistence to soil pH. Figure 7-3 depicts the effect of lime on atrazine persistence in no-tilled soils. In addition, they reported percent weed control was less under no-tillage conditions and was related to the amount of atrazine remaining in the soil (Figure 7-4).

Similar results for chloro-*s*-triazine persistence under no-tillage culture have been reported. Parochetti (1978) found barley yields reduced only at relatively high rates (6.6–8.8 kg/ha) and only in some years. Burnside and Wicks (1980) reported atrazine carry-over in reduced and no-till cropping systems common in arid to semi-arid conditions was less of a problem than under conventionally tilled systems.

Surface plant residue obviously intercepts herbicide spray as it is applied, but the total effect of this residue as it affects weed control is not clearly understood. Lowder and Weber (1979) found that significantly more atrazine could be removed from fresh oat residue than dried corn residue. Also, one 10-cm rain removed more atrazine than did four 2.5-cm rains. Recovery of atrazine was approximately

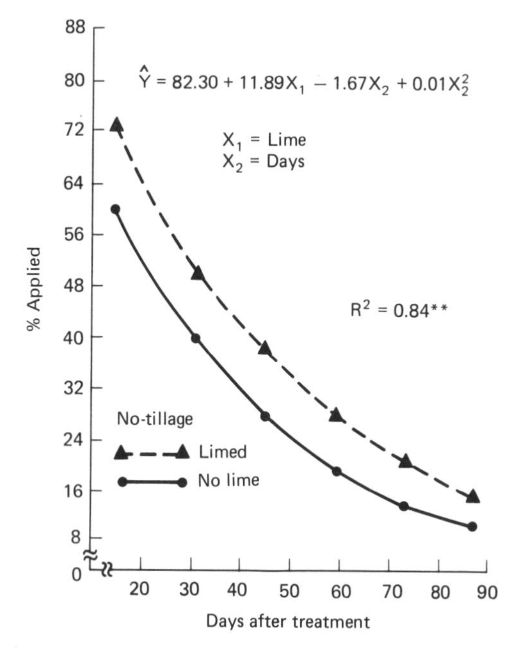

Figure 7-3. The effect of lime on the amount of atrazine remaining in the soil over time under no-tillage (from Kells *et al.*, 1980).

75–85 percent of that applied, but as the time interval between application and extraction increased, the amount of atrazine recovered was reduced. This reflects the importance of rainfall being needed relatively soon after application to move herbicides off plant residue and from the soil surface into the soil to be in close proximity to germinating weed seedlings.

A common practice in some areas of the world is to remove unwanted vegetation by burning the crop residue. Nyffeler and Blair (1978) reported that burnt wheat straw reduced the activity of chlorotoluron and isoproturon on blackgrass (*Alopecurus myosuroides* Huds.). However, Jeffrey and Henard (1970) reported increased weed control in no-till soybeans after burning wheat straw. Responses of herbicides and weeds would be expected to differ in their response to ash, and whether the ash was on the soil surface, as Jeffrey and Henard reported, or incorporated into the soil, as reported by Nyffeler and Blair (1978). Also, the burning process could remove many difficult to control weeds that would not be easily controlled by herbicides.

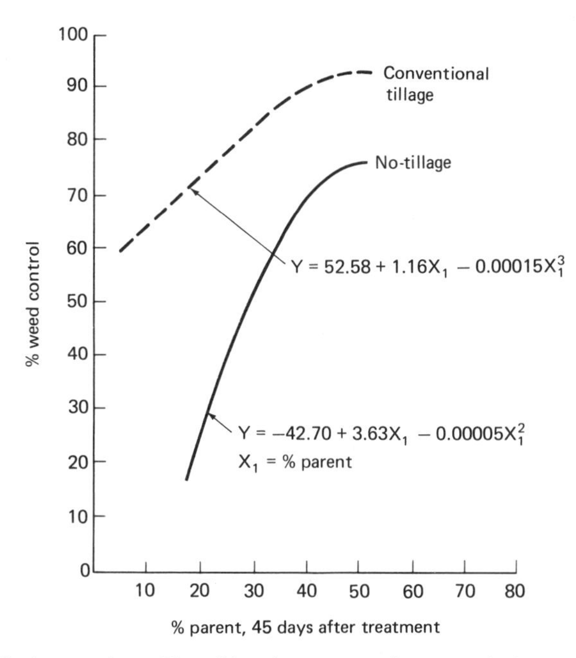

Figure 7-4. Weed control as affected by the amount of parent triazine present in the soil 45 days after treatment under no-tillage and conventional tillage (from Kells *et al.*, 1980).

REFERENCES

Bachthaler, G. 1974. The development of the weed flora after several years' direct drilling in cereal rotations on different soils. *12th Br. Weed Control Conf., Proc.* **3**:1063–1071.

Blevins, R. L., G. W. Thomas, and P. L. Cornelius. 1977. Influence of no-tillage and nitrogen fertilization on certain soil properties after 5 years of continuous corn. *Agron. J.* **69**:383–386.

Burnside, O. C. and G. A. Wicks. 1980. Atrazine carry-over in soil in a reduced tillage crop production system. *Weed Sci.* **28**:661–666.

Chappell, W. E. and L. A. Link. 1977. Evaluation of herbicides in no-tillage production of burley tobacco *(Nicotiana tabacum). Weed Sci.* **25**:511–514.

Cussans, G. W. 1966. Practice of minimum cultivation, the weed problem. *8th Br. Weed Control Conf., Proc.* **3**:884–889.

Elliot, J. G. 1974. Developments in direct drilling in the United Kingdom. *12th Br. Weed Control Conf., Proc.* **3**:1041–1049.

Erbach, D. C. and W. G. Lovely. 1975. Effect of plant residue on herbicide performance in no-tillage corn. *Weed Sci.* **23**:512–515.

Faix, J. J., D. W. Graffis, and C. J. Kaiser. 1979. Conventional and zero-till planted alfalfa with various pesticides. Illinois Agric. Exp. Sta. DSAC 7. Dixon Springs Agricultural Center, Simpson, Illinois, pp. 117–123.

Faix, J. J., G. E. McKibben, and C. J. Kaiser. 1977. Sod suppressants and preemergence herbicides for sod seeded alfalfa. *North Central Weed Control Conf., Proc.* **32**:59–61.

Fenster, C. R. and L. R. Robison. 1970. Herbicides for native sod planted corn. *North Central Weed Control Conf., Proc.* **25**:55.

Griffith, D. R. 1970. Influence of tillage systems on corn production. *North Central Weed Control Conf., Proc.* **25**:23.

Hanson, N. S. 1944. Erosion control in relation to weed eradication. *North Central Weed Control Conf., Proc.* **1**:112–117.

Henard, T. S., H. Andrews, and L. N. Skold. 1969. Effects of herbicides on soybeans and grain sorghum sown in undisturbed small grain stubble. *South. Weed Sci. Soc., Proc.* **22**:184–189.

Herron, J. W. and S. H. Phillips. 1969. Performance of several herbicides in no-tillage corn production. *North Central Weed Control Conf., Proc.* **24**:16.

Herron, J. W., C. E. Rieck, and C. Slack. 1974. No-till renovation of forages with herbicides. *North Central Weed Control Conf., Proc.* **29**:63.

Herron, J. W., L. Thompson, and C. H. Slack. 1973. Evaluation of herbicides used in double-cropped soybeans. *South. Weed Sci. Soc., Proc.* **26**:65.

Hiltboldt, A. E. and G. A. Buchanan. 1977. Influence of soil pH on persistence of atrazine in the field. *Weed Sci.* **25**:515–520.

Jeffrey, L. S. and T. S. Henard. 1970. Effect of row-spacing and stubble burning on weed control in soybeans planted in wheat stubble. *North Central Weed Control Conf., Proc.* **25**:53.

Kells, J. J., C. E. Rieck, and R. L. Blevins. 1979. Simazine activity as affected by two tillage systems. *South. Weed Sci. Soc., Proc.* **32**:47.

Kells, J. J., C. E. Rieck, R. L. Blevins, and W. M. Muir. 1980. Atrazine dissipation as affected by surface pH and tillage. *Weed Sci.* **28**:101–104.

Klingman, G. C. and J. M. Spain. 1965. Chemicalization of sod planting. *South. Weed Sci. Soc., Proc.* **18**:145–152.

Kosovac, Zdravko. 1968. Application of herbicides as an alternative to ploughing and cultivation in sunflower and wheat production. *9th Br. Weed Control Conf., Proc.* **2**:849–854.

Lowder, S. W. and J. B. Weber. 1978. Lime and tillage effects on atrazine efficacy in a clay loam soil. *South. Weed Sci. Soc., Proc.* **31**:73.

Lowder, S. W. and J. B. Weber. 1979. Atrazine retention by crop residues in reduced-tillage systems. *South. Weed Sci. Soc., Proc.* **32**:303–307.

McClure, W. R., S. H. Phillips, and J. W. Herron. 1968. No-tillage experiences in Kentucky. *Proc. 1968 Annual Meeting, Am. Soc. Agric. Eng.*

McKibben, G. E. 1970. Where are we in double cropping in zero-till? *North Central Weed Control Conf., Proc.* **25**:25.

McKibben, G. E. 1979. Tillage comparison, soybeans in corn stalks. Illinois Agric. Exp. Sta. DSAC 7. Dixon Springs Agricultural Center, Simpson, Illinois, pp. 78–80.

Moomaw, R. S. and A. R. Martin. 1976. Weed control in reduced tillage corn planting systems. *North Central Weed Control Conf., Proc.* **31**:47.

Nyffeler, A. and A. M. Blair. 1978. The influence of burnt straw residues or soil compaction on chlortoluron and isoproturon activity. *1978 Br. Crop Protection Conf. – Weeds, Proc.* **1**:113–119.

Parochetti, J. V. 1978. Dissipation of triazines in conventional and no-tillage corn. *Northeastern Weed Sci. Soc., Proc.* **32**:56.

Pavlychenko, T. K. 1944. Optimum depth for germination of weed and grain crop seed in relation to weed control in a growing crop. *North Central Weed Control Conf., Proc.* **1**:106–112.

Peters, E. J. and S. A. Lowance. 1970. Planting pearl millet in sod suppressed with several herbicides. *North Central Weed Control Conf., Proc.* **25**:56.

Peters, E. J. and S. A. Lowance. 1977. Pasture renovation with glyphosate and paraquat. *North Central Weed Control Conf., Proc.* **32**:59.

Pollard, F. and G. W. Cussans. 1976. The influence of tillage on the weed flora of four sites sown to successive crops of spring barley. *1976 Br. Crop Protection Conf. – Weeds, Proc.* **3**:1019–1028.

Roberts, H. A. and P. A. Dawkins. 1967. Effect of cultivation on the numbers of viable weed seeds in soil. *Weed Res.* **7**:290–301.

Roberts, H. A. and P. M. Feast. 1972. Fate of seeds of some annual weeds in different depths of cultivated and undisturbed soil. *Weed Res.* **12**:316–324.

Roberts, H. A. and P. M. Feast. 1973. Changes in the numbers of viable weed seeds in soil under different regimes. *Weed Res.* **13**:298–303.

Ross, M. A. and J. L. Williams. 1969. An analysis of the stale seedbed technique for Indiana soybean production. *North Central Weed Control Conf., Proc.* **24**:13–14.

Sarpe, N. 1974. The results from seven years' research into minimal and zero tillage techniques for maize in Romania. *12th Br. Weed Control Conf., Proc.* **1**:371–378.

Scharbau, W. 1968. Recent developments in the use of herbicides to replace cultivations in some European arable crop situations. *9th Br. Weed Control Conf., Proc.* **3**:1306–1317.

Schnappinger, M. G., C. P. Trapp, J. M. Boyd, and S. W. Pruss. 1977. Soil pH and triazine activity in no-tillage corn as affected by nitrogen and lime applications. *Northeastern Weed Sci. Soc., Proc.* **31**:116.

Shear, G. M. 1965. The role of herbicides in no-tillage crop production. *South. Weed Sci. Soc., Proc.* **18**:28–34.

Slack, C. H. 1973. Influence of tillage on the persistence of chloro-*s*-triazine herbicides. Master's Thesis, University of Kentucky, Lexington, Kentucky.

Slack, C. H., R. L. Blevins, and C. E. Rieck. 1978. Effect of soil pH and tillage on persistence of simazine. *Weed Sci.* **26**:145–148.

Wiese, A. F. 1956. Preliminary research in developing a chemical fallow farming system. *North Central Weed Control Conf., Proc.* **13**:34.

Wiese, A. F. and T. J. Army. 1960. Chemical fallow in dryland farming. *North Central Weed Control Conf., Proc.* **17**:19–20.

Williams, J. L. and M. A. Ross. 1970. Tillage influence on weedy vegetation. *North Central Weed Control Conf., Proc.* **25**:23.

Witt, W. W. 1981. Weed control in no-tillage. In: *No-tillage Research: Research Reports and Reviews.* R. E. Phillips, G. W. Thomas, and R. L. Blevins, Editors. University of Kentucky.

Worsham, A. D. 1970. No-tillage soybeans in mowed and combined wheat stubble. *South. Weed Sci. Soc., Proc.* **23**:87.

8
Other Pests in No-Tillage and Their Control

Shirley H. Phillips
Associate Director of Extension
College of Agriculture, University of Kentucky

INSECTS

Innate doubts associated with any major change of accepted cultural practices tend to surface during the early years of adoption. This is especially true when research data and grower experience are lacking or at best limited. Major concerns in the 1960s were raised relative to mulch buildup and the anticipated potential increase in severity of insect infestations and disease problems. Many of the recommendations on pest control were based on deep coverage of crop residues by plowing and clean cultivation. No-tillage reversed this concept and the idea of leaving mulch and crop residues on the soil surface appeared to present an ominous situation. In addition, germinating and growing plants to be desiccated with herbicides were present in the field and could also serve as host plants for pests.

Early no-tillage systems were more likely to involve corn planted into grass and legumes. Many of these sods were established for several decades with normal insect increases associated with long-term, undisturbed pastures. The soil insect complex, especially wireworms and white grubs, provided with an abundance of food in the stable pasture situation. These insects rapidly turned to feed on newly seeded, germinating and developing plants when the existing vegetation was eliminated with herbicides. The ability to control insects has developed satisfactorily to permit growing of crops in no-tillage systems.

Raney (1981), Kentucky, summarized insect problems in no-tillage crop production in an effective manner. No-tillage can create increased damage from seed pests because of slower increases in soil temperatures, higher moisture and slower germination in temperate zones. A more conducive microenvironment is established because of a lack of mechanical disturbance, higher moisture and less temperature variation. It is difficult to predict the probability or amount of damage from soil pests, the least understood of all economic insect species.

An increase in above-ground insects has also been observed. Gregory (1974) indicated that grasses, grass/legume mixtures, hay and small grains all favor reproduction of soil and above-ground corn insect pests. He also speculated that one could expect a higher potential for damage from the true armyworm, cutworms and European and southwestern corn borers. Further, the elimination of the natural hosts of certain insects by herbicides can force them to the cultivated crop.

Insect management under no-tillage is challenging because of problems in pesticide application for soil pests. Certain insecticides will reduce stands because of their phytotoxicity. However, in-furrow applications of some systemic insecticides have shown promise in moderating soil insect stress and certain foliar pests when used at high rates/ha. The importance of timely rescue-type treatments for above-ground pests is the same as that for conventional corn.

Insect problems in no-tillage systems may also be reduced by the following: (1) possible increase in predator and/or parasite activity; (2) resistant and/or more tolerant varieties; (3) rotational sequences offered by multi-cropping; (4) improved fertilization and increased seeding rates with narrower rows; (5) the compensation mechanism inherent in the cultivated plant; and (6) utilizing Integrated Pest Management strategies.

Soil Insects

The soil-insect complex poses the greatest problem in insect control. These insects can rarely be detected or identified except by extracting them from the soil, which makes biological studies very difficult. Soil-insect pests of corn, in order of observed and/or anticipated importance, are cutworms, wireworms, rootworms, root aphids, white grubs, seed corn beetle, seed corn maggot and grape colaspis.

The black cutworm, *Agrotis ipsilon* (Hugnagel), is potentially the most serious pest of corn seedlings under no-tillage. Musick and Petty (1974) observed that the black cutworm in Ohio attacked approximately 15 percent of the plants in no-tillage corn fields, whereas in adjacent, conventionally tilled fields, only one percent were attacked. Bushing and Turpin (1976) studied oviposition preferences of black cutworm moths and found them attracted to high plant density and low growth form plants. Weedy or grassy areas of fields provide focal points for cutworm infestation and the presence of corn or soybean field debris can increase the potential for cutworm problems.

Sechriest and York (1967) found that while regrowth of damaged corn plants varied with stage of development, plants which are cut off rarely produce ears. Sechriest (1967) found that diazinon, dyfonate, dylox, dursban and thimet showed potential usefulness as insecticides for control of the black cutworm. Pelleted bait formulations are effective if leaf damage can be detected during early larval stages (Sechriest and Sherrod, 1977). Growers must watch their fields closely and be ready to apply baits or directed sprays when plant damage is first noted.

Wireworms of numerous genera damage seeds by depletion of endosperm stores resulting in abortion or weakened seedlings. They also cause damage by boring into the underground portions of the stem near the crown. Musick and Petty (1974) hypothesized that no-tillage corn following sod could have a higher probability of wireworm problems since eggs are laid in the soil near the roots of grasses. Also, since larvae move deeper into the soil as it dries, and since no-tillage increases soil moisture, wireworms could be a serious pest.

The performance of in-furrow application of cyclodiene, organophosphate or carbamate insecticides or seed treatment methods have produced inconsistent results in wireworm control. Apple *et al.* (1958) reported potential advantage for wireworm control using seed treatments for corn. Wireworm suppression is possible with in-furrow application of either Furadan (carbofuran) or Counter (terbufos).

Further, diazinon seed treatments have shown promise where wireworm pressure is light to moderate. The cost of broadcast incorporation of the currently available chemicals to protect corn from wireworm attack is not economically feasible and is impractical under the no-tillage practice.

Corn rootworms (northern, western and southern) have been considered as the most serious insect pest of corn. Musick and Petty (1974) observed that no-tillage caused an approximate four-fold increase in the number of rootworm eggs when compared to conventional systems. If corn stalks or crop residues were removed from the no-tillage system, the egg count was similar to the conventional system. However, Musick and Collins (1971) found that survival of eggs and larvae was reduced under no-tillage. It took approximately four times more eggs in no-tillage to give a mean root rating equal to the conventional system. According to this Ohio study, larval populations were reduced by 33 percent for each 25-cm increment the new corn row was displaced from the old row.

Apple (1961) tested selected granular insecticides against the northern corn rootworm, *Diabrotica longicornis* (Say). These materials were drilled into the soil above the seed. He found that untreated plots had 96.6 percent lodging and 16 percent barren stalks, while treated plots averaged only two percent unproductive stalks. Musick (1975) reported that three years' research in Ohio with the *D. longicornis* (Say) showed that the liquid fertilizer-insecticide combinations do not control rootworm larvae as effectively as banding either spray or granular formulations of insecticides over the row at planting time. There are good insecticidal controls available for larval rootworms in row or band applications. Also, adult control at pollination time should be practiced in order to ensure good kernel set if populations warrant.

Root aphids cause considerable damage to both no-tillage and conventional corn in several regions. Characteristic field appearance is non-uniformity of stand, missing plants and seedlings showing tip necrosis and wilting. Corn field ants and efficient symbionts of these aphids. It is quite common to find aphid establishment concurrent with seed germination. In early spring, aphids are moved to perennial weeds by the corn field ant. Kentucky studies have indicated a selective preference for the genera "Rumex" and "Plantago," with curly dock being the most preferred weed host. This plant might serve as an indicator for predicting anticipated aphid activity in the spring prior to planting. Visual inspection of fields for signs of ant activity is also important. In light to moderate ant/aphid activity in no-till, one finds a preponderance of mounds in the furrow row. Heavy infestations are characterized by generalized ant activity. Excavated plants have aphid colonies established on the roots.

In-furrow placement of selected organophosphates and carbamate insecticides have proven effective. Gregory (1974) reported substantial yield increases in no-tillage corn when comparing the best treatment to the untreated plots.

No-tillage following old sods is highly suspect for white grubs. The lack of physical and mechanical disturbance and failure to expose grubs to parasites and predators enhances their population. Economic injury levels need to be established. In-furrow or band applications of insecticides are of questionable value for white grub control.

If seed germination is delayed in no-tillage systems due to lower soil temperatures, it can extend the vulnerability time for seed pests. As a result, adults of the seed corn beetle destroy seed endosperm stores. Surface trash and decaying organic matter from crop residues provide an ideal situation for reproduction and development of the seed corn maggot. Larval development can proceed at temperatures of 10°C and higher. Early Organophosphate seed treatments have provided satisfactory control for both pests.

Kentucky's Integrated Pest Management program (IPM) utilizes scouts to take soil samples in corn to determine possible soil insect stress. As a result, many participants have decreased or eliminated the use of soil insecticides, especially in conventional corn or in no-till where no soil insects are found.

IPM in Kentucky is a concept that may utilize a variety of control tactics in an overall management strategy that better ensures favorable economic, ecological and sociological consequences. Scouts are hired by farmers to scout their crops for various insects, weeds and diseases and this information is utilized to make a management decision. Economic thresholds are utilized for various above-ground pests in making management decisions whether the field be planted no-till or by conventional means. Raney (1981) noted that this concept is a valid management tool and farmers should utilize its concepts to make timely management decisions.

Above-Ground Insects

The intensity and frequency of above-ground insect pests are increased in no-tillage. Above-ground pests, in order of problem potential, are armyworms, European and southwestern corn borers, stalk borers, flea beetles (Stewart's wilt), corn leaf aphid (MDMV), leafhopper (MCDV), slugs, mice, moles and birds.

Cool, wet springs often precede damage by true armyworms, *Pseudaletia unipuncta* (Haworth). In this kind of weather, larval development is enhanced and the parasite/predator complex is less effective. Moths are active in Kentucky by April 1 and in Ohio by April 10. Females lay eggs on lower leaves of many grasses and cereal plants on which the young larvae feed. After the existing vegetation is killed, the larvae move to and feed readily upon young corn plants. Early season armyworm attacks can markedly affect yields, with reductions related to date and degree of damage.

Musick (1973) noted that if less than 40 percent of the leaf area was removed and the whorl was intact, yields were reduced by approximately 10 percent. However, if plants were pruned to immediately above the growing point on June 13 and June 27, yields were reduced 20 percent and 30 percent, respectively. From this study he concluded it to be economically feasible to control army-worms if 25 percent of the plants show significant damage.

Musick and Suttle (1973) found that if carbofuran was applied in late May at the Maximum registered rate of 28 g ai/300 m of row, maximum suppression of armyworm occurred. Although carbofuran applied at 28 g ai/300 m of row was less effective than higher rates in suppression of severe armyworm damage to early-May planted corn, it significantly reduced the amount of damage by the armyworm over no treatment at all. Bhirud and Pitre (1972) found that when carbofuran or disulfoton were applied in row to corn at planting, there was apparent uniform distribution of the toxicant into both sides of the plant indicating an equal distribution and absorption of insecticide throughout the root zone.

Armyworm control is possible either with properly timed foliar applications or the in-furrow use of carbofuran at planting. The success of foliar applications is dependent upon early detection and timing.

Common stalk borer females lay eggs in weedy and grassy areas. The larvae bore into a wide range of host plants. Stalk borers are limited to field peripheries in conventional corn but occur throughout no-till fields. Since this borer moves from killed hosts and readily bores into young plants, contact surface sprays are ineffective. Systemic insecticides appear to be only moderately effective at present.

Since the European corn borer overwinters as a mature larva in stalks of corn, populations can be reduced by plowing under crop residues. This is not possible with no-tillage. Also, under no-tillage the corn plant will be more attractive to second-generation moths.

Stalk tunneling by second- and third-generation European corn borers results in lodging problems. Also, ear feeding may reduce yields and increases the likelihood of disease. There are numerous literature citations on foliar insecticide control of the European corn borer. Studies by Berry *et al.* (1972) and Edwards and Berry (1972) screened a large number of insecticides for European corn borer efficacy, both as foliar and as soil applications, and established carbofuran as the most effective corn borer control.

The southwestern corn borer, "Diatraea grandiosella" (Dyar), was originally described in 1911 from specimens taken from corn in Guadalajara, Mexico. Davis *et al.* (1974) thoroughly described the biology of this new pest. Even though southwestern corn borer has not received as much publicity as the European corn borer in the northern corn belt states, it is one of the major insect factors limiting corn production in parts of Kentucky. Since 1966 the southwestern corn borer has become established in 49 counties, which, according to Koepper (1974), represent 73 percent of the field corn acreage. Girdling damage by second-generation larvae interferes with mechanical harvesting and increases the incidence of ear rots. Larvae overwinter at the base of the stalk. Successful overwintering of higher population numbers is common where tillage is minimized. The value of early planting to decrease southwestern corn borer damage was established by Walton and Bieberdorf (1948).

Non-chemical control can be effective in reducing southwestern corn borer infestions; however, chemical control remains an integral part of most management programs. Henderson and Davis (1970) indicated that corn cannot always be planted early enough to prevent damage by second- and third-generation borer. Presently, resistant varieties are being tested by commercial seed companies, but this method of control is not yet available to growers.

Conventional insecticide applications will control the southwestern corn borer. Arbuthnot and Walton (1954) and Rolston (1955) first demonstrated effective control of this insect pest. Rolston obtained partial control of southwestern corn borers on whorl-stage corn but did not recommend treatment of older corn because of the excessive amount of spray required.

Since timing of conventional insecticide applications is difficult for optimum effectiveness and repeated insurance applications are costly, the concept of a systemic insecticide was introduced. Hensley *et al.* (1964) were the first to achieve effective results with a systemic insecticide (AC47470).

Davis *et al.* (1974) found in-furrow treatments with carbofuran at planting to be generally as effective for control of second generation southwestern corn borers as were the foliar applications of carbofuran timed to coincide with moth emergence. However, the in-furrow treatments with carbofuran were not as effective in reducing the infestations of the European corn borer as were the foliar treatments at mid-season or at harvest.

Gregory (1974) compared the efficacy and practicality of at-planting systemic applications, systemic plus foliar, and foliar applications for control of the southwestern corn borer. The results of these tests indicated the most promising method of chemical control to be row application of a systemic material.

Insect vectors of disease are important when considered in the context of no-tillage operations. The reservoir of the pathogens if increased and conditions are good for vector survival.

Perhaps the most serious non-insect problems encountered in no-tillage cropping have been from field mice, birds and slugs. The nature of the no-tillage practice itself encourages the mouse problem. Mice feed on young stems of orchardgrass and fescue. When this vegetation is killed with herbicides, their food supply is destroyed. As a result, the mice feed on the corn as long as the kernel remains attached to the roots. When the kernel decomposes, the mice discontinue feeding on corn. Thus, the mice are pests for a period of three weeks after planting. A few mice will remain in the field following that time, feeding on regrowth of grasses and weeds not killed by the herbicides, thereby giving a beneficial service.

Beasley and McKibben (1975) reported control of field mice with Mesurol and Zinc phosphide. Zinc phosphide was formulated as a 2 percent bait and applied broadcast and in-furrow. Mesurol (R) was applied as a seed treatment. Untreated control plots showed 25.3 percent damage and produced 7,100 kg/ha of corn. Zinc phosphide plots averaged 0.4 percent damage and 8,400 kg/ha; whereas Mesurol plots averaged 4 percent damage and 8,000 kg/ha. Presently, zinc phosphide is not registered for use in no-tillage corn. Mesurol (R) is registered for use as a bird repellent but is available as a state label (24-C) for reduction of rodent damage in some areas.

Many believe that no-tillage farming has resulted in increased crop damage because exposed seeds attract birds. Blackbirds, starlings and crows are common pest species. Materials commonly used as "bird repellents" have varied but can be grouped into two principal categories: repellents and toxicants. Chemicals in the first group include coal tar, turpentine, kerosene, cresol, wood oil, thiram, captan, graphite, copper oxalate and red lead oxide. Common toxic materials are lindane, strychnine, phorate, disulfoton, dieldrin and isotox. Perhaps the most promising commercially available repellent is Mesurol (R). Raney (1976), using Mesurol (R) resulted in a stand of 95 plants/3,050 cm of row, compared to an untreated control having a stand of 50 plants/3,050 cm of row.

Barry (1969) found that serious slug problems may occur in no-till, especially where moisture conditions persist in cornfields with surface debris. Musick (1972) found one application of phorate applied to the soils surface at 1 kg ai/ha as a broadcast spray, greatly reduced population numbers of the gray garden slug, "Derocerus reticulatum" (Miller). Musick and Petty (1974) observed that slug damage in Ohio fields appear to be most severe to corn following alfalfa-clover sods or to heavily manured fields. Also, warm, wet springs are conducive to high populations. Mesurol (R) is perhaps the most promising slug control chemical for the no-tillage farmer.

Insects are not regarded as an insurmountable problem in no-tillage or multi-cropping systems. Insect stresses are, however, different from those encountered in conventional production. Even so, insects may be controlled with proper management techniques. These methods, coupled with regular field observations, should greatly reduce the chance for serious problems.

In summary, pest problems will be greater in no-tillage, requiring the grower to be more observant and aware of needed technology to prevent insect problems. The insect situation has not been as severe as early predictions and has not reduced the interest and growth of no-tillage crop production.

DISEASES

The most pleasant surprise with no-tillage cropping has been the lack of increase in economically important disease problems for many of the economically important diseases. Theoretically, one might have expected increased disease pressure in no-tillage, as clean cultivation and plowing under of crop residues has often been recommended as a major method to combat disease.

Research findings comparing disease incidence among various tillage systems indicate a low frequency of increased disease in reduced tillage. Several of these reports are noted later in this section. There is still a need for continued monitoring, however, as the acreage of no-tillage crops is small in many sections of the world and the history of no-tillage in most fields is less than 25 years.

Three criteria must be met simultaneously in order for a disease to occur: a host susceptible to the disease, a virulent pathogen, and an environment conducive to development of the disease. The absence of any one of the three conditions will eliminate disease development.

Many pathologists report that more pathogenic organisms are present in no-tillage than in conventional tillage, but that there is also a higher incidence of parasitic organisms that are capable of combatting these pathogens. In conventional tillage, even with deep plowing of crop residues, enough residue may escape being covered that there is sufficient inoculum present to result in disease. Only small amounts of virulent pathogens need be present with susceptible host plants and favorable environmental conditions for disease to develop in conventional tillage. Rotation of crop species is a common practice in no-tillage that is likely to suppress disease problems to a normal level.

Boosalis and Doupnik (1976) stated that with the evidence available it is not practical to advocate the elimination of reduced tillage practices for the control of plant diseases, since great benefits are derived from reductions in cost of crop production, soil compaction, wind erosion, soil temperature, root injury due to post emergence, cultivation and reduced soil moisture evaporation.

Two standard procedures recommended in reduced tillage systems are the use of resistant varieties where available and the avoidance of a monoculture. Northern corn leaf blight, yellow leaf blight and southern corn leaf blight are examples of diseases of corn for which resistant hybrids are available for planting in reduced tillage regimens. This approach should prove successful until new races of the pathogen develop that are capable of infecting previously resistant hybrids. Monoculture is the continuous cropping of the same host in successive seasons. Where a two- or three-year crop rotation is practiced, the debris of a given crop that may harbor pathogens has generally deteriorated before the same crop is planted again. A double crop system in which three crops are rotated in a two-year period is quite popular in Kentucky. This means that in a given crop rotation (i.e., corn-wheat-soybeans), each crop is planted once within the two years, thus allowing any crop debris to decompose prior to planting the same crop again. Debris of the other two crops in the system generally does not harbor pathogens detrimental to the third crop (Williams, 1974).

Reports comparing disease incidence in conventional with that in reduced tillage systems may at times appear to be contradictory, but much of the conflict may be due to different years, geographical locations, varieties, environments, cropping rotation and other cultural manipulations. More comparative studies are needed in which comparisons are made side by side in the same year with the same variety while minimizing other variables.

Observations and reports of corn, sorghum, soybeans and small grain diseases in reduced tillage systems are explained below.

Corn

Price (1972) states that diseases and yields are not adversely affected by reduced cultivation techniques. Research in eastern Canada indicated little or no yield reduction from diseases in reduced and no-tillage systems.

Seedling diseases such as northern corn leaf blight, yellow leaf blight, "Helminthosporium" leaf spot and rust were observed in West Virginia (Elliot, 1976) to be equally injurious under no-till, minimal-till and conventionally tilled fields.

In New York a portion of the yield increase in conventional tillage was attributed to less eye spot (*Kabatiella*) (Boothroyd, 1976). There was no appreciable difference in stalk rots. The application of Dithane M-45 to decrease leaf spotting was more effective in conventional tillage than in no-tillage.

Earlier and more severe infection of southern, northern and yellow corn leaf blights has been observed in Indiana with surface residue systems (Griffith *et al.*, 1977). However, it should be noted that resistant hybrids to these diseases are available where desired. Nonplowed fields were also observed to have an earlier onset of corn anthracnose and more severe brown spot and "Diplodia" and "Gibberella" ear rots than in comparable plowed fields (Griffith *et al.*, 1977).

A higher incidence of Maize Chlorotic Dwarf Virus (MCDV) and Maize Dwarf Mosaic Virus (MDMV) and in general greater disease severity was associated with lower yields in no-till versus conventional tillage systems in Georgia (All *et al.*, 1977).

It was noted that since 1971 there has been a greater development of gray leaf spot on no-tillage in Virginia (Roane *et al.*, 1974). (An illustration of what effect one disease has on others is that MDMV plants were more susceptible to gray leaf spot and stalk rot, and breakage was greater where gray leaf spot was a problem). A report from Tennessee (Hilty, 1976) supports the observations reported in Virginia claiming that gray leaf spot is becoming increasingly severe in several counties where minimum tillage is practiced.

No appreciable differences in the incidence of gray leaf spot in conventional and no-tillage systems have been observed in eastern Kentucky. In 1977 in western Kentucky, conventional and no-till corn were planted in adjacent fields. A greater incidence of Stewart's wilt, MDMV and MCDV diseases was observed in conventional tillage. The no-tillage showed an increased early season antracnose, had a sparser stand and had a greater incidence of northern corn leaf blight. Phillips and Young (1973) listed 26 diseases of corn and found an increased disease incidence as a result of no-tillage practices in two of the diseases, antracnose and yellow leaf blight.

Aflatoxin problems in corn, especially in southeast United States, were quite severe in 1977. Much of the problem was attributed to the summer drought. In reduced tillage systems, moisture is conserved, resulting in less plant stress and perhaps fewer aflatoxin problems.

Sorghum

Where crop debris is allowed to remain intact on the soil surface, diseases expected to be of special concern would include root and stalk rots and bacterial and fungal blights. Although the expectation seems logical, several reports have suggested that it is unwarranted. The ecofallow method (reduced tillage) tested for a three-year period in Nebraska showed lower incidence of stalk rots and higher yields than conventional tillage method (Boosalis and Doupnik, 1976). Stalk rot incidence in conventional tillage was 39 percent, in minimum tillage 23 percent and in no-till 11 percent. After six consecutive years in a wheat-sorghum ecofallow system, the incidence of sorghum stalk rot was reduced 30 percent. Price (1972) also notes that sorghum diseases and yields were not adversely affected by reduced cultivation techniques.

Soybeans

Soybean varieties resistant to foliar and root rot diseases are not as plentiful as are resistant corn hybrids. Therefore, since resistant varieties planted in no-till have limitations, crop rotation becomes very important for soybean disease control in no-till. Planting directly into wheat stubble by a no-till planter was not as satisfactory in Arkansas (Collins *et al.*, 1975) as it had proven to be in northern states.

Seedling diseases, predominately "Fusarium" and "Rhizoctónia," appeared as major problem in double cropped soybeans. However, soybean stem canker in Georgia (Phillips, 1976) was found in all areas and was not associated with any particular variety or cultural practice. "Phytophthora" root rot, on the other hand, was found only in north Georgia and was usually found in no-till plantings of the Ransom variety. Less "Cylindrocladium" black rot on soybeans was observed in Virginia (Garren and Porter, 1976) when soybeans were planted after winter wheat than when planted after fallow. In *No-Tillage Farming* (Phillips and Young, 1973), bacterial blight, bacterial pustule, wildfire, anthracnose and sclerotial blight disease problems were reported to increase in no-tillage while "Phytophthora," "Rhizoctonia," "Fusarium" root rot and stem rot diseases were decreased.

Several soybean diseases increase in severity when soybeans follow soybeans in conventional tillage. An assumption could be made that the diseases would be even more severe with no-tillage. Observations made in Kentucky fail to support that assumption, however. Brown spot, pod and stem blight and anthracnose disease were more prominent in a conventional tillage field than in the adjacent no-till field. Both fields had identical field histories. Lodging was found to be just as extensive or more extensive in the no-till field. In a small plot experiment in western Kentucky, brown spot disease occurred much earlier and was more severe in plots where a soybean-fallow-soybean rotation was used than where a soybean-wheat-soybean was used, regardless of the tillage system employed. Observations of disease problems were no greater, even perhaps less severe, in double cropped no-till soybeans than in conventionally planted soybeans. Two factors are perhaps quite important in influencing these results. The first is that, in Kentucky, most conventionally planted soybean fields are planted earlier than double cropped no-till soybeans and the chances are good that the field was in soybeans the previous year. The double cropped no-till planted soybeans were likely to have been planted much later after wheat harvest, and more than likely had a two-year crop rotation of corn-wheat-soybeans. More than likely the time of planting and the crop rotation had more influence on disease incidence than the tillage method used.

Wheat

Wheat, a nearly universal crop, has had more worldwide attention to tillage systems than any other single crop. In England (Lockhart *et al.*, 1975) Take-All disease appeared in a much lower incidence in no-till than in conventional deep plow or shallow plow. In the third and subsequent years of wheat monoculture in Switzerland (Vez, 1975), direct sowing (no-tillage) increased grain yields and decreased attack by fungal diseases. Eyespot problems in Germany (Schwerdtle, 1971) during two to three years of continuous cropping with winter wheat were less severe on direct sown (no-tillage) than on conventionally cultivated plots. After the third year, Take-All and, to a lesser extent, "Cercospora" disease problem reports were similar to results reported for eyespot disease (Debruck, 1971).

In Montana (Dubbs, 1976), although "Cephalosporium" root rot and cheatgrass became problems after three crop years in the winter wheat-fallow rotation, they were not present in the three year rotation of winter wheat-barley-fallow. Four years of field data showed conclusively, under Pullman, Washington conditions, that burial of straw residue by moldboard plowing versus a stubble mulch method of residue management favors better wheat growth and increased yield but at the same time favors more "Cercosporella" root rot (Cook, 1975). Conversely, another Montana report (Aase and Siddoway, 1976), suggested that the advantages of no-till planting in stubble were masked because standing stubble harbored disease (no specific diseases mentioned) which attacked wheat in the early growth stages, thereby decreasing yields.

Where "Fusarium" root rot was suspected in Oregon (Ramig *et al.*, 1976), the disease severity was similar in all tillage treatments of fall plow, spring plow and no-till. Another Oregon report (Rydrych, 1975) found that a marked reduction in root rot disease in no-till culture (whether or not the same root rot disease is uncertain) occurred while increased disease problems were involved in plow and stubble mulch culture. According to a Washington report (Cook and Rovira, 1976), Take-All disease is fostered by wheat monoculture but may be significantly reduced by fallow soil or crop rotation. Minimum tillage reduced the Take-All disease (Cook and Rovira, 1976; Rydrych, 1975).

Hood (1965), Jealotts' Hill Research Station in England, reported reduced problems with Take-All (*Gaeumannoymces graminis* var. *tritici*) disease in wheat no-tillage, when compared to plowed, cultivated and drilled tillage.

The disease incidence in the third crop of wheat after permanent pasture in the sprayed/direct drilled plots were 2.6 for Take-All (*Gaeumannoymces graminis* var. *tritici*), 26.0 for Eyespot (*Cerosperella herpotrichoides*), 617 for Sharp Eyespot (*Rhizoctonia solani*) and 15.1 for Brown Foot Rot (*Fusarium* spp.). In the ploughed/cultivated drilled plots, Take-All percent infected tillers of total tillers was 20.4, Eyespot 38.2, Sharp Eyespot 5.7, and Brown Foot Rot 9.7. These comparisons show a highly significant reduction in Take-All in sprayed/direct drilled plots compared to conventional seeding methods and no significant difference in Eyespot, Sharp Eyespot and Brown Foot Rot.

Summary

In no-tillage, anthracnose and yellow leaf blight, diseases of corn are increased and Take-All in wheat is substantially less severe, demonstrating the wide variation that may occur between crops and pathogens in no-tillage. Improved moisture regimens promote rapid growth and healthier plants, which reduces severity of disease in many instances. Serious disease problems have yet to develop over a wide area of the world with no-tillage production. No-tillage allows more flexibility in rotation of crop species which can reduce disease risks. Growers should practice the use of resistant crop varieties, avoid monoculture, use good practices that enhance plant growth and follow normal disease control programs adaptable to no-tillage culture.

ACKNOWLEDGMENT

Dr. Richard Stuckey, Associate Professor of Plant Pathology, University of Kentucky, contributed the major content on disease problems in the *Diseases* section.

Drs. Harley Raney and Lee Townsend, Extension Professor of Entomology and Assistant Extension Professor of Entomology, respectively, University of Kentucky, provided the major content for the *Insect* section.

REFERENCES

Aase, J. K. and F. H. Siddoway. 1976. Management and use of precipitation and solar energy for crop production in the northern great plain. USDA/CRIS. Progress Report.

All, J. N., C. W. Kuhn, R. N. Gallaher, M. D. Jellum, and R. S. Hussey. 1977. Influence of no-tillage-cropping, carbofuran and hybrid resistance on dynamics of maize chlorotic dwarf and maize dwarf mosaic diseases of corn. *J. Econ. Entomol.* **70**(*2*):221–225.

Apple, J. W. 1961. Appraisal of insecticidal granules in the row against damage by the northern corn rootworm. *J. Econ. Entomol.* **54**:833–836.

Apple, J. W., F. E. Strong, and E. M. Raffensperger. 1958. Efficacy of insecticidal seed treatments against wireworms on lima beans and corn. *J. Econ. Entomol.* **51**:690–692.

Arbuthnot, K. D. and R. R. Walton. 1954. Insecticides for control of the southwestern corn borer. *J. Econ. Entomol.* **47**:707–708.

Barry, B. D. 1969. Evaluation of chemicals for control of slugs on field corn in Ohio. *J. Econ. Entomol.* **62**:1277–1279.

Beasley, L. E. and G. E. McKibben. 1975. Controls for mouse damage to no-till corn. Illinois Agric. Exp. Sta. DSAC **3**:96.

Berry, E. C., J. E. Campbell, C. R. Edwards, J. A. Harding, W. G. Lovely, and G. W. McWhorter. 1972. Further field tests of chemicals for control of European corn borer. *J. Econ. Entomol.* **65**:1113–1116.

Bhirud, K. M. and H. N. Pitre. 1972. Bioactivity of carbofuran and disulfoton in corn in greenhouse tests, particularly in relation to leaf positions on the plant. *J. Econ. Entomol.* **65**:1183–1184.

Boothroyd, C. W. 1976. Diseases of Field Corn. USDA/CRIS. Progress Report.

Boosalis, M. G. and B. Doupnik, Jr. 1976. *Management of Crop Diseases in Reduced Tillage Systems. Bulletin of the Entomological Society of America* **22**(*3*):300–302.

Busching, M. K. and F. T. Turpin. 1976. Oviposition preferences of black cutworm moths among various crop plants, weeds and plant debris. *J. Econ. Entomol.* **69**:587–590.

Collins, F. C., C. E. Caviness, and H. D. Scott. 1975. Maximizing soybean yields in double-cropping systems. USDA/CRIS. Progress Report.

Cook, R. J. 1975. Management of soilborne diseases of wheat. USDA/CRIS. Progress Report.

Cook, R. J. and A. D. Rovira. 1976. The role of bacteria in the biological control of *Gaeumannomyces graminis* by suppressive soils. *Soil Biol. Biochem.* **8**(*4*):269–273.

Davis, F. M., D. A. Henderson, and T. G. Oswalt. 1974. Field studies with carbofuran for control of the southwestern corn borer. Mississippi Agric. and Forestry Exp. Sta. Bull. No. 807.

Debruck, J. 1971. Soil cultivation and direct-sowing on weak pseudovergley para brown earth. Landwirtschaftliche Forschung. *Sonderheft* **26**(*1*):230–244.

Dubbs, A. L. 1976. Methods of production of small grains and forages. USDA/CRIS. Progress Report.

Edwards, C. R. and E. C. Berry. 1972. Evaluation of five systemic insecticides for control of the European corn borer. *J. Econ. Entomol.* **65**:1129–1132.

Elliot, E. S. 1976. Diseases of Forage Crops. United States Department of Agriculture/Current Research Information Service. Progress Report.

Garren, K. H. and D. M. Porter. 1976. Peanut disease investigations. USDA/CRIS. Progress Report.

Gregory, W. W. 1974. No-tillage corn insect pests of Kentucky . . . A five year study. *Proc. Nat. No-Tillage Research Conf.* **1**:46–58.

Gregory, W. W. and H. G. Raney. 1980. Pests and their control – insect management. In: *No-Tillage Research: Research Reports and Reviews.* R. E. Phillips, G. W. Thomas, and R. L. Blevins, Editors. College of Agriculture, University of Kentucky.

Griffith, D. R., J. V. Mannering, and W. C. Moldenhauer. 1977. Conservation tillage in the eastern corn belt. *J. Soil and Water Cons.* **32**(*1*):20–28.

Henderson, C. A. and F. M. Davis. 1970. Four insecticides tested in the field for control of *Diatraea grandiosella. J. Econ. Entomol.* **63**:1495–1497.

Hensley, S. D., W. Machado, and D. R. Melville. 1964. Control of the southwestern corn borer with an experimental systemic insecticide. *J. Econ. Entomol.* **57**:1011.

Hilty, J. W. 1976. Viruses and mycoplasma-like organisms causing diseases of corn and sorghum. USDA/CRIS. Progress Report.

Hood, A. E. M. 1965. Ploughless farming using "Gramoxone." *Outlook on Agric.* **4**:286–294.

Koepper, J. M. 1974. Kentucky agricultural statistics 1973. Kentucky Crop and Livestock Rep. Serv.

Lockhart, D. A. S., V. A. F. Heppel, and J. C. Holmes. 1975. Take-All (*Gaeumannomyces graminis* (Sacc.) Arx and Oliver) incidence in continuous barley growing and effect of tillage method. *Bulletin OEPP* **5**(*4*):375–383.

Musick, G. J. 1972. Efficacy of phorate for control of slugs in field corn. *J. Econ. Entomol.* **65**:220–222.

Musick, G. J. 1973. Control of armyworm in no-tillage corn. *Ohio Report* **58**(*2*):42–45.

Musick, G. J. 1975. Best corn rootworm control with banded insecticides. *Ohio Report* **60**(*1*):3–5.

Musick, G. J. and D. L. Collins. 1971. Northern corn rootworm affected by tillage. *Ohio Report* **56**(*6*):88–91.

Musick, G. J. and H. B. Petty. 1974. *Insect Control in Conservation Tillage Systems. Conservation Tillage . . . A Handbook for Farmers.* Soil Conservation Society of America, 52 pp.

Musick, G. J. and P. J. Suttle. 1973. Suppression of armyworm damage to no-tillage corn with granular carbofuran. *Ohio Report* **66**:735–737.

Phillips, D. V. 1976. Prevalence, control and pathogenesis of soybean diseases. USDA/CRIS. Progress Report.

Phillips, S. H. and H. M. Young, Jr. 1973. *No-Tillage Farming.* Reiman Associates, Milwaukee, Wisconsin.

Price, V. J. 1972. Minimum tillage: looks like a winner. *Soil Conser.* **38**(*3*): 43-45.

Ramig, R. E., R. R. Allmaras, and C. L. Douglas. 1976. Tillage and residue practices for erosion control in the northwest. USDA/CRIS. Progress Report.

Raney, H. G. 1976. Personal communication (Extension Professor of Entomology). University of Kentucky, Lexington.

Raney, H. G. 1981. Kentucky Integrated Pest Management. U.S.D.A. Annual Evaluation Report, pp. 1-31.

Roane, C. W., R. L. Harrison, and C. F. Genter. 1974. Observations on gray leaf spot of maize in Virginia. *Plant Disease Reporter* **38**(*5*):456-459.

Rolston, L. H. 1955. Insecticides tested against the southwestern corn borer (Lepidoptera:Pyralidae). *J. Kans. Entomol. Soc.* **28**:109-114.

Rydrych, D. J. 1975. Cultural, chemical, and biological weed control in the Columbia basin. USDA/CRIS. Progress Report.

Schwerdtle, F. 1971. Trials on direct-sowing methods in comparison with conventional cultivation of various crops with particular regard to the weed flora. *KTBL-Berichte uber Landtechnik No. 149,* 139 pp.

Sechriest, R. E. 1967. Studies on black cutworm control. *Proc. of N.C.B.-E.S.A.* **22**:89-93.

Sechriest, R. E. and D. W. Sherrod. 1977. Pelleted bait for control of black cutworm in corn. *J. Econ. Entomol.* **70**:699-700.

Sechriest, R. E. and A. C. York. 1967. Evaluating artificial infestations of black cutworm. **60**:923-925.

Walton, R. R. and G. A. Bieberdorf. 1948. The southwestern corn borer and its control. Oklahoma Agric. Expt. Sta. Bull. No. B-321.

Williams, A. S. 1974. Personal communication (Extension Professor of Plant Pathology). College of Agriculture, University of Kentucky, Lexington.

Vez, A. 1975. Trial to diminish the effects of wheat monoculture by applying various cultivation practices. *Revue Suisse d'Agriculture* 7(*5*):153-158.

9
Changes in Soil Properties Under No-Tillage

Robert L. Blevins
Professor of Agronomy
University of Kentucky

and

M. Scott Smith
Associate Professor of Agronomy
University of Kentucky

and

Grant W. Thomas
Professor of Agronomy
University of Kentucky

When farmers shift from a system of agricultural production that includes numerous tillage operations to a reduced or no-tillage system it is reasonable to show concern about how this change in soil management may affect soil properties. For any crop production system to be widely accepted and used it must maintain the physical properties of the soil, and allow for replacement of nutrient removal and other losses. It must also maintain a soil environment favorable for the numerous necessary biological reactions. The microbial activity of the soil is largely determined by the chemical and physical properties of the soil. For example, the placement and amount of organic material directly influences biological activity, as it also does the chemical and physical properties.

In the no-tillage system, by leaving the residues on the soil surface with no mechanical mixing of residues and added amendments, we have a soil with contrasting soil biological, chemical and physical properties as compared to plowed soils. A comparison of no-tillage to more conventional tillage will include differences in the microbial environment, number and activity of soil microorganisms, tillage effect on soil animals, decomposition of organic matter, nitrogen transformations, chemical properties, influence of mulches on soil physical properties and effect of tillage on soil density and porosity.

THE MICROBIAL ENVIRONMENT

The number and activity of the microorganisms in soil is determined by the chemical and physical characteristics of the soil environment. Since many biologically significant properties differ between no-tillage and tilled soils, effects of tillage on soil microbes and on the transformations of organic matter and mineral nutrients which they carry out would be expected. The most important soil factors from the microbe's point of view may be the difference in the distribution and quantity of organic material in no-tillage soils and the differences in moisture regime.

Differences in moisture and temperature resulting from no-tillage practices affect microbial activity. No-tillage, or any other tillage system which leaves a mulch of plant residue on the surface, will generally increase soil moisture content. Under dry conditions this would favor microbial activity in no-tillage soils relative to plowed soils, but it also suggests that a no-tillage soil is more prone to become water-saturated and anaerobic subsequent to rainfall or irrigation of moderate intensity. This could affect transformations of N fertilizer. Thermal insulation by the surface mulch could likewise have effects which are dependent on climatic conditions. In the spring no-till soils are often cooler and this would reduce microbial activity, but an equally important and a stimulatory effect is the moderation of temperature fluctuations in soils with a surface mulch. As a rule of thumb, microbial activity doubles for every 10°C increase in temperature. Since reported temperature differences between no-tillage and plowed soils are less than this at most times, the effects of temperature may not be large except at limited times during the year.

Acidification of the surface of no-till soils could inhibit the activity of certain sensitive microbes – nitrifiers, for example. Differences in soil structure and the physical disruption due to plowing undoubtedly have important indirect effects on the biochemical environment.

In studying the biological effects of different tillage methods, the soil microbiologist is thus presented with the interesting situation of soils under identical climates and crops, with similar pedogenic and mineralogical properties, but which can differ greatly in their biochemical characteristics.

Number and Activity of Soil Microorganisms

It has been realized for many years that the method of tillage and treatment of crop residues can affect microbial populations in the soil. Early investigations, though, were conducted on stubble-mulched or sub-tilled soils rather than on no-till. Although this system does involve some mechanical disruption of the soil, it resembles no-tillage in that much of the crop residue remains on the surface. In Maryland, Dawson (1945) originally reported no consistent differences in populations of bacteria or fungi when crop residues were either left on the soil surface as a mulch, disked in, or plowed. However, he later (Dawson *et al.*, 1948) observed differences in the distribution of microbes in sub-tilled compared to plowed soil. The largest effect was the greater populations of fungi, bacteria and actinomycetes in the surface 0–2.5 cm of the sub-tilled soil. Populations tended to be not significantly different or slightly higher in the plowed soil from 2.5–15 cm. Olson and McCalla (1960) in Nebraska measured greater respiration rates in the surface (0–2.5 cm) of stubble-mulched or sub-tilled compared to plowed soils, but there were no effects of tillage for samples from deeper in the profile. However, their laboratory assays were conducted on mixed, frozen soil samples and so may not be applicable to *in situ* microbial respiration. In another Nebraska study, fungal populations were greater by one third to one half in the surface of sub-tilled compared to plowed soils, but there were no significant or consistent differences in populations of bacteria and actinomycetes (Norstadt and McCalla, 1969). It was also found in this study that removal of residues from the surface resulted in lower microbial numbers. Doran (1980a) has also shown that the size of the microbial population is related to the amount of residue applied in minimum till soils.

Direct-drilling of small grains as practiced in Great Britain is also comparable to no-till in that there is minimal mechanical disturbance of the soil. However, formation of an organic matter mulch is often deliberately prevented by burning off crop residues. Plate counts of fungi but not of bacteria were higher in 0–5-cm samples of direct-drilled relative to plowed soil. The direct-drilled soil also evolved more CO_2 (Barber and Standell, 1977). Lynch and Panting (1980a) estimated microbial biomass in 0–5-cm samples of direct-drilled and plowed soils. Biomass was greater in direct-drilled on five of six sampling dates, with ratios of direct-drilled to plowed soil biomass in the range of 0.8–1.7. The significance of the surface mulch, or lack of it, was demonstrated in a later study (Lynch and Panting, 1980b). When residues were chopped and left on the surface of the direct-drilled soil, there was approximately twice as much microbial biomass from 0–5 cm as there was when residues were removed by burning. Some data in this study appear to be at odds with the previously cited work by the same authors; on four sampling dates direct-drilled soil biomass was less than plowed. This inconsistency was not noted nor was there an explanation offered.

Only recently have investigations on the microbial ecology of true no-till soils appeared in the literature, though numerous workers are currently active in this area. Therefore, it may be premature to offer more than a few simple generalizations about microbial numbers and activity in no-tillage compared to tilled soils. Also, it would be surprising if the apparent effects of no-tillage management did not vary with climate, cropping system and history, soil type, and perhaps most important, the methodology used by the soil microbiologist. The effects that are consistently observed include a concentration of microbes at the surface of no-till soils, small increases in no-till or insignificant differences with regard to total number of microbes in the profile, and somewhat greater apparent microbial activity in no-till relative to plowed soils during most of the year.

The concentration of microbes at the surface of no-till soils is illustrated in Figure 9-1. These plate counts of aerobic microorganisms were obtained from a Kentucky soil which had been in continuous corn with no-tillage or conventional tillage (plowing and disking) for

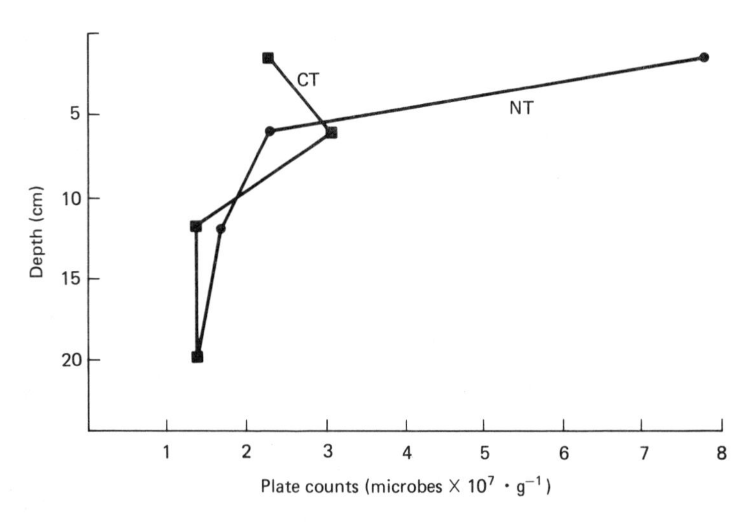

Figure 9-1. Plate counts of aerobic soil bacteria at varying depths in no-till (NT) and conventionally-tilled (CT) Maury silt loam soil. (From Smith, 1979)

nine years (Blevins *et al.*, 1977). During this period of time there has been considerable accretion of organic material at the surface of the no-till soil. This includes both a surface mulch of only partially decomposed plant materials and a lower layer, ranging from 1–5 cm in thickness, consisting of highly decomposed, amorphous material resembling the organic horizon of a forest soil. It is in this latter zone where microbial activity is particularly high. In plowed soil where crop residues are mixed through the plow layer, rather than left on the surface, microbial counts are generally observed, as is the case in Figure 9-1, to be relatively constant with depth within the plow layer. Since microbial growth in soil is believed to be most often limited by the supply of energy source (organic matter for a large majority of microbes) it is expected that the distribution of microbes would be well-related to the distribution of decomposable organic material.

Doran (1980b) conducted a thorough study of microbial populations and biochemical characteristics of soil samples from seven different locations throughout the U.S. where no-tillage versus conventional tillage experiments had been in progress for three to ten years. These experiments included both continuous corn and wheat-fallow cropping systems and also varied in previous cropping history and, of course, climate. Nevertheless, similar effects of no-tillage on microbial distribution were generally observed. A summary of his results is presented in Table 9-1. All categories of microorganisms enumerated were higher in no-tillage than conventional tillage soil from 0–7.5-cm depth. This relationship was usually reversed, at least for aerobic organisms, at the 7.5–15-cm depth, and summing the counts from 0–15 cm showed little net difference in the aerobic microbial populations. There were significant total increases in numbers of facultative anaerobes and denitrifiers in the no-till soils. This led Doran (1980b) to the conclusion that the environment under no-tillage is "less oxidative," implying more anaerobic microbial processes, such as denitrification and fermentation. In West Virginia, Fairchild and Staley (1979) observed no difference in populations of bacteria, fungi, actinomycetes or extractable ATP (a useful index of microbial biomass or activity) in 0–30-cm samples of no-till and plowed soils. They did not sample smaller depth increments. Likewise, Mitchell *et al.* (1980) in Florida detected no significant effect of tillage on total microbial populations, but gave no information on the depth of their samples. Wacha and Tiffany (1979) identified 1,658 fungal isolates from no-till and plowed plots in Iowa (0–15-cm samples). There were no remarkable effects of tillage on the fungal species, with regard to either diversity or frequency of species

Table 9-1. Ratio of microbial populations in no-till and conventional-till soil; average of six locations. (Reproduced with the permission of John W. Doran, Soil Scientist, Agricultural Research Service, U.S. Department of Agriculture, Lincoln, Nebraska, from *Soil Science Society of America Journal* 44:765–771 [1980].)

	RATIO OF POPULATIONS (NT/CT) BY SOIL DEPTH (cm)		
MICROBIAL GROUP	0–7.5	7.5–15	0–15
Total aerobic microbes	1.35[1]	0.71[1]	1.03
Fungi	1.57[1]	0.76[1]	1.18
Actinomyces	1.14	0.98	1.08
Bacteria (aerobic)	1.41[1]	0.68[1]	1.03
Nitrifiers	1.13	0.56[1]	0.87
Denitrifiers	2.70	1.92	1.99

[1] Significant effect of tillage at 0.05 level.

isolated. Tillage significantly affected the frequency of occurrence of only 2 of the 59 species identified.

The effects of tillage on endomycorrhizal fungi have also been considered. These form a symbiotic relationship with the roots of most higher plants including grasses and legumes. Kruckelmann (1975) counted fewer spores of endomycorrhizal fungi in no-till than in plowed soil. However, tillage by rotary hoe decreased spore counts below that in no-till. When additional straw residue was applied to the soil, the relationship between tillage treatments was the same but the differences were accentuated. In contrast, Mitchell *et al.* (1980) observed greater numbers of spores in conventional till soil, but differences were significant only after plant harvest. Root infection, however, was greater in no-till soil. The effect of tillage on the development of mycorrhizae is a question that warrants further investigation. The same can be said of another symbiotic soil microbe-*Rhizobium*.

The enumeration of soil microbes, as described above, is of interest in terms of microbial ecology but has limited value for predicting tillage effects on specific microbial processes or even the overall level of microbial activity. One reason for this is that enumeration procedures can exclude certain active organisms but include spores and inactive forms. Another approach to measuring microbial activity is to observe the rate of a microbial reaction or group of reactions in the soil. Soil enzyme assays have been used to characterize the biochemical effects of tillage. Klein and Koths (1980) reported higher urease, protease and acid phosphatase activity in the surface 0–10 cm of a Connecticut soil under no-till compared to conventional tillage.

In Germany no-tillage has been observed to result in usually greater soil dehydrogenase activity (Greilich and Klimanek, 1976). The dehydrogenase assay involves the reduction of a tetrazolium dye and has been proposed to be an index of the respiratory activity of soil microbes. Doran (1980b) has used this assay and a soil phosphatase assay to characterize tillage effects at seven U.S. experimental sites. For 0–7.5 cm, samples no-till soils had greater dehydrogenase activity than plowed soils at five of seven locations, and no-till and higher phosphatase activity at six of seven sites. Below 7.5 cm there was no consistent effect of tillage on enzyme activities.

A popular and more direct method of estimating microbial activity in soils is the measurement of CO_2 production. This has yet to be widely applied to tillage studies. Measurements of CO_2 flux from the soil surface made late in the season in Kentucky indicate significantly more activity in no-till than in plowed soil (Rice and Smith, 1982a). This relationship does not hold for the entire year; immediately after plowing CO_2 flux from conventionally tilled soils is greater. Staley and Fairchild (1979) observed no significant effect of tillage on CO_2 concentrations in the soil profile.

Tillage Effects on Soil Animals

Among the most dramatic biological consequences of tillage are those observed for the soil fauna. Perhaps Charles Darwin (1907) was the first to note that in the absence of the plow, many of this instrument's activities are assumed by soil animals, earthworms particularly. These organisms can, like a plow, accelerate organic matter decomposition by breaking it apart and mixing it into the soil. Burrowing and mixing alters soil structure. Some nematodes and insects can cause plant diseases.

Ehlers (1975) measured twice as many earthworm channels in no-till as in plowed soil. He suggested that these channels are particularly effective in the transmission of water and account for the greater rates of water infiltration usually observed with no-till soil. McCalla (1958) reported that stubble-mulched soils had higher populations of earthworms than plowed soil. In a Connecticut study, earthworm populations were 3.4 times greater under no-tillage as compared to conventional tillage (Bertsch, 1982).

Thomas (1978) observed somewhat greater numbers of total nematodes in no-till plots than in plowed. Yet Mitchell *et al.* (1980) found no effect of tillage on numbers of plant parasitic nematodes. In another study there were fewer plant parasitic nematodes in no-till, but more total nematodes (Corbett and Webb, 1970). Dispersal of nematodes is greatly accelerated by transport of soil from one field to another. Since this occurs during tillage operations, it is reasonable to suggest that the spread of many nematode diseases would be restricted with no-tillage.

Plowing is generally believed to decrease both the species diversity and population size of most other types of soil fauna including arthropods and enchytraeids (potworms), a possible exception being molluscs, particularly slugs (Kevan, 1968; Wallwork, 1976). It has also been our unfortunate personal observation that a larger member of the soil fauna, moles, are more active in no-till plots than in contiguous plowed areas.

The detrimental effect of plowing on soil animals is probably due primarily to the creation of a less desirable habitat. The increased fluctuations in temperature and moisture in tilled land may be quite significant. Elimination of large pores and channels at the soil surface would restrict the activity of some species. Many soil animals are surface feeders and incorporation of plant residue could make their food supply less accessible. The direct physical action of the plow and the resulting abrasion can also result in significant mortality (Wallwork, 1976).

Decomposition of Organic Matter

The fact that plowing increases the rate of organic matter loss from soils is clearly established (Giddens, 1957). Plowing has several effects which accelerate decomposition processes. Disruption of soil aggregates apparently increases microbial access to organic compounds which are otherwise protected by physical entrapment (Rovira and Greacen, 1957). Tillage will to some extent break apart large pieces of plant residue, thereby increasing the surface area available for microbial attack. Mixing of plant material with soil may result in a greater or more uniform initial inoculation with saprophytic microbes. Immediately after plowing, aeration and physical conditions in the soil surface are favorable for microbial activity. Finally, incorporation of organic residues into the soil usually places them in a constantly moist environment, while residues on the soil surface are subject to desiccation – an effective method of preserving organic materials.

It is almost always observed that soils managed by no-tillage for extended periods of time have higher organic matter contents than if they are plowed (Lal, 1976; Blevins *et al.*, 1977). As is the case for microbial populations, the differences in the distribution of organic matter are often more striking than the differences in total quantity in the profile. Thus, in Kentucky we observed 50 percent more organic carbon from 0–5 cm in no-till soil, but only small differences between tillage treatments below 5 cm (Blevins *et al.*, 1977). Since comparable masses of plant material are synthesized annually in no-till and plowed soils, yet no-till soils have more organic matter, we can only conclude that microbial decomposition of organic matter is slower in no-till soils.

In light of this seemingly inescapable conclusion, direct evidence may appear superfluous. There is, however, sufficient data available to support this conclusion. Figure 9-2 shows the results of a Kentucky study in which corn stalks or rye were enclosed in open-mesh nylon bags, then either placed with the plant residue on the surface of the no-till soil or buried 7–15 cm deep as the conventional till soil was being plowed. After recovery from the soil, the organic matter remaining in the bag was measured. Within one week, half of the rye in the conventionally tilled plots was gone, but approximately four weeks were required to reach the same state of decomposition in the no-till plots. Corn stalks also decomposed more slowly on the

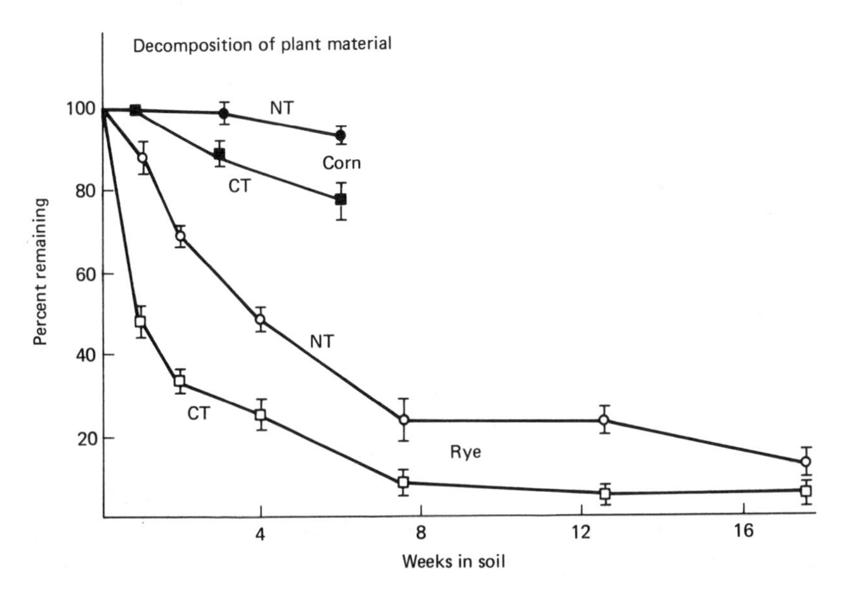

Figure 9-2. Decomposition of plant residues, either corn stalks or rye leaves and stalks, in no-till (NT) and conventionally-tilled (CT) Maury silt loam soil. (From Rice and Smith, 1982a)

surface of the no-till plots. Similar results have been reported by others, including Hice and Todd (1980) in Georgia. Several earlier studies indicate that plant residues will decompose faster when buried than when left on the surface, regardless of tillage system (Parker, 1962; Brown and Dickey, 1970; Sain and Broadbent, 1977).

The slower rates of organic matter decomposition observed for no-till soils appears to contradict the findings discussed above showing that no-till soils usually have greater or equal microbial populations and activity. Since organic matter decomposition powers microbial activity, it would be surprising if there was no correlation between these parameters. This contradiction may be explained by the fact that most biochemical studies of no-till soils have ignored seasonal variations. It is probable that microbial activity and population growth is greater in a conventional till soil immediately after plowing, a time when organic matter breakdown is rapid. Following this flush of activity, the more favorable moisture status and greater quantity of substrate remaining in the no-till soil could lead to greater microbial activity. Most measurements of microbial numbers and activity reported in the literature appear to have been made during the latter phase.

An interesting and perhaps agronomically important issue related to organic matter decomposition is the accumulation in soil of microbiologically produced phytotoxic organic compounds. Under certain environmental conditions, particularly when soils are cool and damp, germination and seedling growth are sometimes reduced where there is an accumulation of surface mulch. This could lead to poor stands in no-till soils. Although a large number of factors could account for this effect, phytotoxicity is a frequently proposed explanation. Norstadt and McCalla (1968, 1969) observed greater populations of *Penicillium urticae* in subtilled than in plowed soil. Populations were smaller where residue was removed from the surface of the sub-tilled soil. This fungus produces the phytotoxin, patulin. Patulin was detected in field samples of residue and soil in this study. Cochran *et al.* (1977) suggest that phytotoxin production may be responsible for the poor stands sometimes obtained with no-till wheat cropping. Plant toxicity was demonstrated in water extracts from mulches composed of wheat straw, lentil straw or bluegrass. Toxicity was greatest following periods of cool, wet weather. In more recent work phytotoxicity was detected by bioassay in extracts of wheat straw incubated for long periods of time in cool, wet soil (Elliott *et al.*, 1981). However, data from the same study on wheat growth in pot culture suggest that phytotoxicity was not a primary cause of reduced yields in mulched soils; at least under the conditions of the study, immobilization of fertilizer nitrogen was a more important effect.

Mineralization and Immobilization of Nitrogen

The shape of the N response curve for no-till soils appears to differ from that for plowed soils in a fairly consistent way in the temperate regions. No-till soils yield less at sub-optimal rates of N but equal or greater yields are attained at higher N rates (see Chapter 5). Although non-biological processes such as leaching of NO_3^- (Thomas *et al.*, 1973; Tyler and Thomas, 1977) may be involved, transformations of soil N by microorganisms must certainly be considered if we are to account for the effects of tillage on the fate of N fertilizer. Higher rates of immobilization and lower rates of N mineralization in no-till soils may explain why yields with no-till are relatively less at low rates of applied N. Since organic matter decomposition is slower without plowing, it would be expected that conversion of organic N to plant-available inorganic N would be slower in no-till. Also it is reasonable to predict that immobilization of added N could be significant with no-till practice because N fertilizer is often placed on the soil surface where there is an accumulation of decomposable organic material. The mulch-forming material in no-till soils, usually wheat straw or corn stalks, is likely to have a high C/N ratio, and this will favor immobilization. Though this is plausible, the available information on tillage effects on mineralization and immobilization are not so clear, so it is not yet possible to be certain of the significance of these processes.

Indirect evidence led Dowdell and Cannell (1975) to the conclusion that decreased rates of N mineralization in direct-drilled soils, and not denitrification or leaching, account for the lower concentrations of NO_3^- usually observed. Burford *et al.* (1976) also suggested that mineralization was less in direct-drilled than plowed soil. Soils cropped by no-tillage for several years have more *potentially* mineralizable N than plowed soils (Doran, 1980b). This can be taken as an indication that less mineralization has actually occurred *in situ* in the no-till soils. Doran also states that there is an increased potential for immobilization of surface-applied N fertilizer in no-till. Powlson (1980) attempted to directly measure mineralization in the field in tilled and non-tilled plots. The soils studied were low in organic matter and residues were removed by burning; consequently, increases in mineral N were small for all treatments. However, the apparent mineralization during the first two weeks after tillage was three to four times greater in plowed than in untilled plots. It was suggested that the effect of tillage on mineralization in these soils was not large enough to account for the greater N fertilizer requirement of direct-drilled wheat, and that other processes such as leaching and denitrification should be considered. Kladivko and Keeney (1981) have also directly measured the effect of tillage on mineralization in the field. A preliminary report of their results suggests that there was somewhat less mineralization with minimum till (Buffalo till planter) than with moldboard or chisel plow, but the differences were neither large nor consistent.

Other studies indicate that there will not always be less mineralization and more immobilization in no-till soils. By measuring corn yields following alfalfa with varying tillage and N treatment, Triplett *et al.* (1979) concluded that "there was no detectable difference in mineralization of N due to tillage." However, since responses to N were of limited magnitude for all treatments it might have been difficult to detect differences in mineralization with only yield measurements. A study by Bennett *et al.* (1975) apparently indicated greater mineralization with no-till than with plowed soil. Mineral N was less in plowed plots in the fall; laboratory mineralization potential was greater for untilled soil; and yields were higher on untilled soil. These are all indirect measurements of mineralization in the field and the observations might be attributed to other factors.

In laboratory experiments by Cochran *et al.* (1980) there was less potential for immobilization of N with surface-applied than with incorporated wheat straw. Similar conclusions were reached by Parker *et al.* (1957) and by Elliott *et al.* (1981). These observations can presumably be explained by the more rapid decomposition of the low-N plant materials when they are incorporated into the soil. The timing of mineralization-immobilization reactions can be more important than their magnitude. Intensive immobilization of N following residue incorporation may persist for only a week or two, to be followed by a net release of N, and so could reduce plant uptake of N less than a more prolonged though less severe period of immobilization caused by surface-applied residues.

Using ^{15}N-labeled fertilizer, Dowdell and Crees (1980) found no difference in apparent N immobilization in direct-drilled and plowed soils; nor was there a tillage effect on plant recovery of ^{15}N. If immobilization of N fertilizer is in fact a problem in no-till soils, there may be a practical solution. Placement of N fertilizer below the surface organic matter should minimize the potential for microbial assimilation of N. With surface-applied wheat straw in a greenhouse pot experiment, greater wheat yields were obtained with low levels of incorporated N than with the same rate of surface-applied N (Elliott *et al.*, 1981). Laboratory results by Cochran *et al.* (1980) also indicate that in mulched soils there is a potential for immobilization only in the top few cm of soil. In field trials, no-till corn responded more to injected anhydrous ammonia or injected N solution than to surface-applied N solution, urea, or ammonium nitrate (Mengel, 1980). Several recent, unpublished laboratory and field experiments in Kentucky also indicate increased potential for immobilization in no-till, and suggest that recovery of N by the crop is enhanced by sub-surface placement.

Effects of Tillage on Nitrification and Denitrification

Nitrification and denitrification are additional microbial processes for which an effect of tillage has been indicated. As is the case for mineralization – immobilization and leaching, nitrification and denitrification may have a significant effect on the N fertility of the soil. Yet there is insufficient information available at present to generalize about the relative contribution of these various reactions. The observation that no-till soils contain lower NO_3^- concentrations (Thomas *et al.*, 1973) could be accounted for if nitrification rates were depressed or if denitrification rates were enhanced. There is some evidence that this is the case in some no-till soils.

McCalla (1958) observed no significant effect of tillage on the rate of nitrification. However, these assays were conducted in the laboratory with near optimum soil temperature and moisture for all treatments, so the results may not pertain to field conditions, where these variables may have a significant effect. Greilich and Klimanek (1976) did observe slightly higher rates of nitrification in plowed soils. Hoyt *et al.* (1980) concluded that nitrification is inhibited in no-till relative to convention till. This was deduced from the lower ratio of NO_3^- to NH_4^+ in the no-till soil, 1.2 in no-till but 1.8 in plowed. Nitrifier populations, averaged over seven locations, were higher in no-till at the 0–7.5-cm depth, but higher in conventional-till from 7.5–15 cm (Doran, 1980b). Summing the counts for 0–15 cm, there was a slightly lower population of NH_4^+ oxidizers in the no-till soils. Thus, there is some evidence that nitrification may be slower in no-till soils, but this is clearly not established as a general effect of tillage. One explanation for this effect, where it does occur, could be the acidification of the no-till soil surface (see section on changes in chemical properties in this chapter). Nitrifiers are generally considered to be highly sensitive to acidity (Dancer *et al.*, 1973). Our experiments at Kentucky measuring N transformations in intact soil columns and in the field (Rice and Smith, 1982a) lead to a different conclusion about the effect of tillage on nitrification. In this case, nitrification was often slower in the tilled soils than in the no-till. This was probably caused by moisture limitations and the more rapid drying of the tilled soil by surface evaporation.

The microbial conversion of NO_3^- to N gases, or denitrification, can waste significant quantities of N fertilizer in wet anaerobic soils. Some of the properties of no-till soils suggest that denitrification losses can be larger than in plowed soils. There is often more organic matter, the energy source for denitrification, at least at the surface of no-till soil. Early in the growing season after plowing, a conventional till soil is less compact and so perhaps better aerated. Most significant, no-till soils are normally wetter and so less water is required to attain water-saturated, anaerobic conditions. On the other hand, it could be argued that the lower NO_3^- concentrations in no-till soils limit denitrification, that the lower pH at the no-till surface inhibits denitrification, or that buried plant residues are more likely to result in anaerobic conditions than are residues on the surface.

Counts of denitrifying bacteria provide some indication of the potential for denitrification but it should be remembered that there may be little correlation between actual denitrification rates and numbers of denitrifiers. This is because the population of denitrifiers can be maintained or increased by aerobic growth as well as by denitrification. Doran (1980b) found the average denitrifier population to be seven times greater in the surface 0–7.5 cm of no-till soils compared to plowed soils. In contrast to most other microbial categories counted, denitrifiers were also more numerous from 7.5–15 cm in the no-till soils. In another study on the effect of mulching on soil microbes, Doran (1980a) reported that denitrifier populations were increased by as much as 44-fold in the mulched soils. This was attributed to the soil being wetter and having more carbon substrate. McCalla (1958) suggested that denitrifiers may be more numerous at the surface of sub-tilled soil than plowed soil. Staley and Fairchild (1979) detected no effect of tillage on denitrifier populations.

Methods for direct measurement of soil denitrification are not sufficiently developed for reliable determination of rates under varying tillage treatments in the field (Focht, 1978). Some measurements of one of the denitrification products, N_2O, have been made in direct drilled versus plowed experiments. The flux of N_2O from the surface of direct drilled soil was three to five times greater than from plowed soil (Burford *et al.*, 1977; Burford *et al.*, 1981). Unfortunately, there are other mechanisms of N_2O generation (Bremner and Blackmer, 1979; Smith and Zimmerman, 1981) and N_2 not N_2O is usually the major product of denitrification. Thus, greater production of N_2O does not unequivocally demonstrate greater rates of denitrification.

Using a laboratory assay of denitrifying enzyme activity we have found that no-till soil consistently has greater activity than conventional till (Rice and Smith, 1982b). We have also used intact soil cores from no-till and conventional till soil to directly measure N gas production. Soil cores are treated with acetylene gas, which under the proper conditions arrests the sequence of denitrification reactions at the N_2O stage; i.e., N_2O is the only product (Balderston *et al.*, 1976; Yoshinari and Knowles, 1976; Smith *et al.*, 1978). Following irrigation of the cores, N_2O production is measured by gas chromatography. Representative results are shown in Figure 9-3. No-till

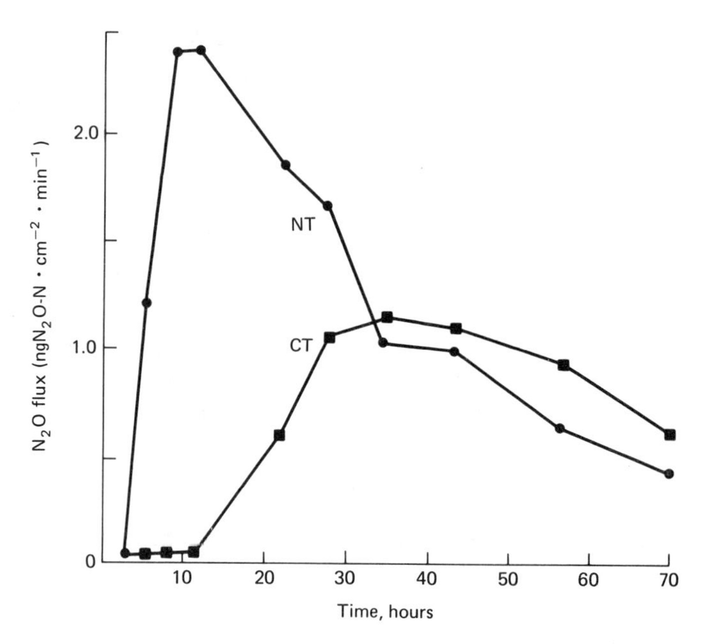

Figure 9-3. Evolution of N gas by intact Maury soil cores from no-till (NT) and conventionally tilled plots (CT) following irrigation with 1.3 cm of water. Production of N_2O by these cores is a direct measure of soil denitrification. (From Rice and Smith, 1982b)

soils show a more rapid and a greater denitrifying response than conventional till soils. The no-till soils we have studied are well-drained; though denitrification rates are apparently greater with no-till, we believe denitrification losses to be large only in very wet seasons on these soils. However, increased denitrification under no-till may be an important factor limiting the application of this practice to less well-drained sites.

SOIL PHYSICAL PROPERTIES

Influence of Mulches on Soil Physical Properties

The addition of mulch material to the soil surface has a positive influence on several soil physical properties such as a higher percentage of macropores, improved soil aggregate stability and increased aggregate size. All of these result in better air and water movement. Additions of a mulch on Alfisols in Nigeria (Lal *et al.*, 1980) produced a positive influence on soil properties by improving the soil structure, soil porosity and water transmission. These beneficial effects from additions of mulches were due in part to a pulverizing and mixing action from soil fauna activity and reduction in raindrop impact. Earthworm activity, for example, not only creates channels that affect water conductivity but produces worm casts that are structurally strong and in effect are stable soil aggregates. All of these factors reduce the potential harmful effect of raindrop impact on soil structure.

Under most climatic and environmental situations, residues returned to the soil surface in no-tillage systems will gradually decompose. Although humus decomposes more slowly than the original raw organic material (residue), it does continue to break down. Continual additions of humus from decomposing residues must be added to the soil system to offset the oxidation of soil humus if a desirable level of soil organic matter is to be maintained or increased. Soil microbes involved in organic matter decomposition produce exudates that serve as binding agents for soil aggregates by holding them together in stable structural units. Therefore, decomposing soil organic materials contribute both to improved soil fertility and better soil physical conditions. Rates of organic matter decay varies with climate, soil type and nature of the original plant material. For example, soils located in warmer, moist locations will decay at a faster rate than cooler or more arid regions. The Van't Hoff rule predicts that for every 10°C increase in soil temperature the rate of biological activity doubles. This helps explain in part why soil organic matter is usually not very high in the tropics even though annual residue production may be enormous.

The rate of decay of plant residues returned to the soil in a no-tillage or reduced tillage system is a key factor in how well the total soil organic matter level is maintained or even increased. Studies by Lucas *et al.* (1977) in Michigan related changes in organic carbon percentage in soils to cropping systems over time. They postulated that soils have a steady-state level or equilibrium level of soil organic matter. This level will change with cropping or tillage systems. For example, they determined that the steady state for a continuous corn system, spring plowed, and a 6 percent slope is 1.4 percent soil carbon and for "no-till" and about 2.3 percent soil carbon (Figure 9-4).

Tillage systems that utilize plant or animal residues promote plant growth through improved soil productivity. Additions of organic matter improves soil structure and supplies required plant nutrients, especially nitrogen. The no-tillage cropping systems return plant residues to the soil surface which adds to the pool of organic material and these additions have a direct and significant influence on reducing soil erosion.

Crops managed for high yields with residues returned can double the level of organic soil carbon over a long period time, if one starts at a low level. The model developed by Lucas *et al.* (1977) in Michigan suggested that increasing soil carbon from 1.0 percent to 2.15 percent would increase corn yield potentials by 25 percent.

Returning corn stover mulch to the soil surface in a no-tillage system conserves soil moisture and improves soil physical properties such as degree of structural development. Improved soil structure and increased porosity result in a higher infiltration capacity. Results from measurements made on a Wooster silt loam soil by Triplett *et al.* (1968) showed increased infiltration rate with increased levels of surface mulch residues. Infiltration after one hour was 1.80, 1.22, 2.34 and 4.39 cm for treatments – plowed-bare, no-tillage bare, no-tillage with 40 percent of the surface covered with residues and

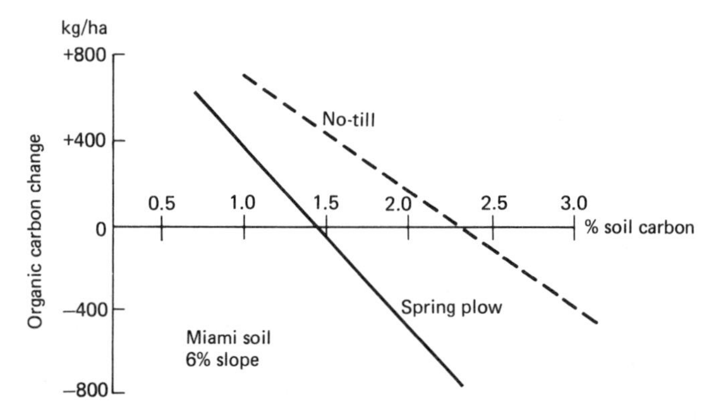

Figure 9-4. Yearly soil organic matter change on a Miami silt loam soil in mid-Michigan in relation to tillage and soil carbon contact. (From Lucas *et al.*, 1977).

no-tillage with 80 percent surface residues coverage, respectively. They concluded that after three years of corn production, the mulch cover had a positive influence on the structural stability of the soils.

Comparisons of infiltration rates of no-tillage and plowed treatments were made by Lal (1976) on Alfisols in western Nigeria. He concluded that no-tillage plots had infiltration rates three times greater than the plowed plots (0.91 cm min^{-1} versus 0.30 cm min^{-1}).

The management of residues in no-tillage or reduced tillage systems emphasizes keeping the residue on or near the soil surface. Under these systems where plant residues are left and medium to high levels of fertility are used, one can expect an increase in organic matter in the soil surface with time. A long-term study at Kentucky (Blevins *et al.*, 1983) is a classic example of how tillage and residue management can modify soil properties. Their findings show that in the 0–5 cm soil depth, organic carbon was approximately twice as high with no-tillage as with conventional tillage. A comparison of tillage methods showed no significant effect of organic carbon at the 5–15- and 15–30-cm depths (Table 9-2). In contrast, a comparison to an adjacent continuous bluegrass sod, conventional tillage corn production after ten years showed a decrease in organic carbon.

The organic matter level in a soil is controlled by the rate of loss through decomposition of the soil organic matter and the rate of gain from formation of soil organic matter during decomposition of plant residues and roots returned to the soil. Barber (1979) in Indiana concluded that cropping practices that remove residue over a period of years following a previous practice where the residue was returned to the soil, the organic matter level would decrease until it comes to

Table 9-2. Comparison of organic carbon content of soils with tillage systems after ten years of corn production. (From Blevins *et al.*, 1983.)

	ORGANIC CARBON (%)		
SOIL DEPTH (cm)	NO-TILLAGE	CONVENTIONAL TILLAGE	UNTREATED BLUEGRASS SOD
0–5	2.71	1.38	3.81
5–15	1.22	1.40	1.73
15–30	0.77	0.96	0.90

an equilibrium with the new practice. (The removal of corn and small grain for silage is an example.) After ten years of continuous corn he reported that the organic matter level decreased to 90 percent of the original level. When residues were returned there was a slight increase (Table 9-3).

The prudent management of residues as a surface mulch not only adds to the pool of decomposing organic material which contributes to the forming of more stable soil structural units but dramatically reduces the impact of falling raindrops. This direct effect of surface mulches of raindrop impact protects the soil aggregate from being broken by sheer energy of the raindrops. The detachment of soil particles from soil structural units is the first step in the mechanisms related to soil erosion. This detachment step is often followed by a crusting of soil surface upon drying. After crusting occurs, the infiltration capacity of a soil is reduced and surface water runoff is increased many-fold. The results are increased runoff, soil erosion losses, sedimentation problems and pollution. These negative conditions can be avoided by utilizing farming systems such as no-tillage or reduced tillage systems that maintain a high percent of residues at the soil surface.

Effect of Tillage on Soil Density and Porosity

An important objective of tillage research is to identify and measure soil conditions resulting from tillage or a lack of tillage and how

Table 9-3. Effect of residue management on percent of soil organic matter. (Reproduced with the permission of S.A. Barber, from *Agronomy Journal* 71:625-627 [1979].)

TREATMENT	SOIL ORGANIC MATTER (%)	
	6TH YEAR	11TH YEAR
Residues removed	2.8	2.8
Residues returned	3.0	3.0
Double residues returned	3.4	3.4
Fallowed for six years, then corn with residues returned	2.6	2.7
L.S.D. (0.05)	0.2	0.2

these changes affect plant growth. In the last two decades there has evolved a greater concern about excessive tillage operations and the detrimental effects resulting from this kind of soil management. The growing of row crops without tillage such as the no-tillage system and reduced tillage systems has caused much concern among soil researchers and farmers about how this practice will effect the overall soil physical properties. For example, soil density and porosity are soil properties that have a profound influence on air and water movement and the potential productivity of a given soil.

Research results on this subject show contrasting results when compared over a wide range of soil types and climatic conditions. In the humid temperate zone of the U.S., results from long-term no-tillage cropping systems show no net change in bulk density resulting from the lack of tillage (Blevins *et al.*, 1983; Shear and Moschler, 1969) (Table 9-4). All of these studies cited above have similar climatic conditions and the soils all have a medium textured soil surface. In contrast, a comparison of the effects of no-tillage with conventional-fall-plow tillage in Minnesota (Gantzer and Blake, 1978) on a clay loam soil (Aquic Argiudoll) produced different results (Table 9-5). This finer-textured soil had significantly greater bulk density in spring and fall on no-tillage-treated plots as compared to conventional tilled plots. The higher bulk density caused lower air-filled porosities of surface soil samples under no-tillage as compared to conventionally tilled plots.

In spite of lower air-filled porosities and higher bulk density values, the no-tillage treatments still had higher volumetric water contents and a higher number of bio-channels. The significant differences between tillage systems for bulk density, porosity and saturated hydraulic conductivity did not occur below a depth of 30 cm. Differences in soil physical parameters resulting from no-tillage

Table 9-4. Soil bulk densities under different tillage systems. (From Blevins *et al.*, 1977; and Shear and Moschler, 1969.)

	BULK DENSITY (g/cm^3)	
TILLAGE SYSTEM	BLEVINS	SHEAR AND MOSCHLER
No-tillage	1.25	1.48
Conventional	1.25	1.43

Table 9-5. Bulk densities of undisturbed soil cores for no-tillage and conventional tillage for fall and spring sampling dates on a Le Seur silty clay loam soil. (Reproduced with the permission of Gantzer and Blake, from *Agronomy Journal* 70:853-857 [1978].)

DEPTH (cm)	NO-TILLAGE	FALL PLOW	STANDARD ERROR	PROB-ABILITY
	(g/cm^3)			
		May 20, 1974		
7.5–15	1.32	1.05	0.02	<0.001
30–37.5	1.29	1.29	0.02	ND
		September 20, 1974		
7.5–15	1.25	1.12	0.03	0.02
30–37.5	1.29	1.27	0.02	ND

compared to conventional tillage were smaller when measured in the fall prior to harvest than those taken in the spring. The fall plowing to start the second year disturbed the conventional tillage plots; therefore the authors were unable to observe the seasonal effect.

At a number of sites in Great Britain it was observed that under direct drilling of grain crops a pronounced increase in bulk density develops at 0–15-cm soil depth (Pidgeon and Soane, 1977; Ellis *et al.*, 1979; Cannell *et al.*, 1980). These differences between tillage systems are related to soil texture and soil structure and other properties that affect soil trafficability. On other soils with poor structure and low organic matter, the mechanical loosening effects from tillage are soon lost by raindrop impact of summer storms and the sealing over of the surface soil layer.

Studies of soil properties of a Latosol influenced by tillage versus direct drilling in Paraná, Brazil (Vieira, 1981) shows higher bulk density for the direct drilled plots (Table 9-6), then the conventionally tilled and slightly lower total porosity. The physical properties of the soil were not influenced below the 15-cm soil depth. The higher macro-porosity for conventional tillage agrees with findings reported by Van Ouwerkerk and Boone (1970); and Gantzer and Blake (1978).

Zero-tillage (no-tillage) experiments in the Netherlands (Van Ouwerkerk and Boone, 1970) showed that untilled soils are usually

Table 9-6. Soil density, porosity, and pore size distribution in a Latosol after four years of direct planting and conventional tillage in Paraná, Brazil. (From Vieira, 1981.)

	SOIL DEPTH	CONVENTIONAL TILLAGE	DIRECT PLANTING
	(cm)		
Soil density (g/cm^3)	15	1.08	1.20
	30	1.14	1.12
	60	0.99	0.97
Total porosity (% volume)	15	62	57
	30	60	60
	60	65	66
Micro-porosity (% volume)	15	36	42
	30	41	40
	60	38	37
Macro-porosity (% volume)	15	26	15
	30	19	20
	60	27	29

denser than plowed soils and mean soil pore space is lower for untilled plots. These findings indicate that no-tillage reduces total pore space and changes pore size distribution, with larger pores disappearing and finer pores predominating. These conditions and results differ from results reported on medium-textured silt loam soils by Triplett *et al.* (1968) in Ohio and Lal (1976) in Nigeria. Their findings suggest an increase in infiltration capacity of soil under no-tillage management and they attribute the increased concentration of organic matter at the surface for improving the infiltration rate. Another explanation for increased infiltration under no-tillage is the fact that continuity of large pores are not disrupted by tillage equipment. The maintenance of pore continuity from surface to the lower soil depth along with increased earthworm activity in some soils (Ehlers, 1975; Barnes and Ellis, 1979) contribute to more favorable soil, air and water movement. Obviously, these different views and research results are real and are influenced by different soils and climatic conditions.

Cannell and Ellis (1977), working with wheat in Great Britain, concluded that no-tillage (direct drilling) compared to conventional tillage resulted in a reduction in total pore space and that a large portion of the remaining pores were filled with water. Similar studies by Douglas *et al.* (1980) showed lower soil total porosity and less volume of transmission pores in direct drilled plots on a clay soil. Despite these findings, an increase in infiltration of water was observed along with an increase in earthworm populations and activity. Thus, soil aeration was improved under no-tillage culture on some soils but poorer aeration was observed on clayey soils during the wet winters.

These changes in soil physical properties are largely determined by the initial soil properties. Generally, no-tillage fields are more resistant to compaction for a given pass with equipment and exhibits more favorable trafficability properties. However, clayey soils when wet, are more easily deformed or reshaped and compaction from wheel traffic of equipment is detrimental. This compaction is not countered by the normal plowing operations that are part of conventional land preparation for seeding a crop. In Britain, where residues are removed, usually by burning, before drilling wheat, the clay soils exhibit a degree of self-mulching (Cannell, 1981) in the upper 2–3 cm. This provides ideal conditions at the end of summer for germination of winter wheat without the straw mulch on the surface.

There is evidence of improved soil properties after repeated direct drillings. Work by Douglas and Goss (1982) showed that aggregate stability of the topsoil was enhanced by direct drilling and shallow tine cultivation. This improved soil structure was related to an increase in soil organic matter. However, in a clay soil that was previously under grass sod, aggregate stability and organic matter content decreased after several years of direct drilled wheat but not as much as treatments after moldboard plowing. These results from Great Britain are similar to observations made in Kentucky (Blevins *et al.*, 1983). After 10 years of continuous corn on plots that had been in a grass sod for 50 years, a slight decrease in organic matter was observed for no-tillage in the soil surface 0–5-cm layer. A 64 percent decrease in organic matter occurred after 10 years of conventionally tilled corn.

Water-stable aggregation is often used as a parameter to evaluate soil structure and is a suitable index of soil resistance to dispersion and compaction. This index influences plant emergence, water intake and soil erosion. Organic matter percentage is often closely related to the degree of soil aggregation. Soil aggregation was studied in Indiana (Mannering *et al.*, 1975) for conventional, chisel, till-plant and no-tillage systems, the averages of four different soils are given in Table 9-7. As tillage increased, soil aggregation decreased. Aggregation was highest in the 0–5-cm zone with no-tillage.

Tillage studies on silty soils in Germany (Ehlers, 1979) characterize the effect of no-tillage to conventional tillage on hydraulic properties of these soils. When tilled, the soil aggregates of these silty soils tend to slake during heavy rains and a silt crust is formed at the surface. A plow-pan is commonly found at the 20–30-cm depth. Under these conditions there was improvement in the aggregate stability of the surface soil under no-tillage. This improvement was attributed to an increase in organic matter. Total porosity was increased by tillage in top 0–10 cm layer. However, due to continuous macro-channel systems connecting the soil surface to the subsoil, infiltration was enhanced by no-tillage management. Unsaturated hydraulic conductivity within the low tension range was higher for no-tillage than the tilled soils. Ehlers (1979) concluded that the continuity of larger pores such as those associated with earthworms was a factor but the lack of disturbance in no-tillage system favored the maintenance of a continuous hydraulic system of finer pores from one soil layer to the next.

CHEMICAL PROPERTIES

No-tillage affects chemical properties of soil as compared to conventional tillage primarily because of three changes in soil treatment.

Table 9-7. Effect of tillage on soil aggregation of four Indiana soils, after five years. (From Mannering *et al.*, 1975.)

	AGGREGATION INDEX[1]	
TILLAGE SYSTEM	0–5 cm	5–15 cm
Conventional	0.35	0.47
Chisel	0.46	0.56
Till-plant	0.47	0.56
No-tillage	0.77	0.70

[1] Determine by a modified Yoder method. Samples were passed through an 8-mm sieve while moist, then air-dried. Stability measurements were made by a wet sieve method.

First, and most important, the soil is not moved, so there is no mixing of soil amendments; second, most fertilizers, lime and other chemicals are applied to the soil surface; and third, the surface mulch changes the evaporation rate of water from the soil surface, especially during periods when the soil is wet and there is a large evaporative demand.

When the soil remains in place year after year and whatever amendments are repeatedly placed on the soil surface, it can be expected that the distribution of non-mobile ions will be far from uniform. In fact, the undisturbed soil acts much like a chromatographic column, with strongly-held ions and/or ions added in large quantities concentrated at the soil surface and the replaced ions moved out. In a very general way, one can expect phosphate and potassium to be concentrated at the surface if fertilizers containing these ions are added regularly. (Data showing this tendency are shown in Table 9-10.) As discussed in Chapter 5, this distribution of nutrients near the soil surface has not been shown to be detrimental to nutrient uptake because high root concentrations occur in the same soil volume and higher water contents are generally observed. The higher water content of the soil promotes a larger diffusion rate of nutrients. Nevertheless, the appropriate depth of sampling to determine the availability of P and K in soils is no longer 15 cm as in the case of a plowed soil, where mixing of applied nutrients is pretty complete within the 0–15-cm depth. Instead, the concentration of both phosphorus and potassium is very high in the 0–5-cm depth and then tails off very sharply. This suggests that a diagnostic depth of soil sampling for no-tillage should be shallower than for plowed soils, perhaps 5–10 cm in depth.

As contrasted to potassium and phosphorus, the levels of calcium and magnesium with depth can vary quite widely. The most important causative factor in the loss of calcium is the amount of nitrogen that is added as fertilizer. Since nitrogen is added as partly or wholely ammonium and the nitrification process produces two hydrogen ions for each nitrite ion formed, the effect can be very large at high nitrogen rates. Table 9-8 shows the calcium concentration with depth for 0, 84, 168 and 336 kg/ha of nitrogen as ammonium nitrate for five and ten years on a Maury silt loam (Blevins *et al.*, 1977, 1983).

Table 9-8. Exchangeable calcium in Maury silt loam under no-tillage for five and ten years. (From Blevins *et al.*, 1977 and 1983.)

	FIVE YEARS			TEN YEARS		
	SOIL DEPTH (cm)			SOIL DEPTH (cm)		
N RATE (kg/ha)	0–5	5–15	15–30	0–5	5–15	15–30
	(meq Ca/100 g)					
0	7.2	7.0	7.1	5.5	6.0	6.6
84	5.4	5.2	5.6	3.8	5.7	6.4
168	5.5	6.8	7.1	2.6	4.7	6.1
336	3.5	5.0	6.8	1.2	1.9	5.2

The Pierre formula for calculating the effect of nitrogen fertilizer on loss of soil calcium predicts that for each unit of nitrogen, 1.8 units of calcium carbonate will be lost. The data in Table 9-8 are somewhat variable but, looking at only the surface 0–5-cm depth, the loss of calcium carbonate per unit of nitrogen added was in the range of 1.5–4.5. The rate declines with higher and higher nitrogen rates, indicating that when much of the calcium is already gone, removal of the rest is less efficient. This calculation assumes that the no-nitrogen plot was losing, on the average, 393 kg/ha of calcium carbonate each year. During the first five years, the loss was 425 kg/ha-year and during the second five years the loss was 358 kg/ha-year.

Using the average of 2.6 units of calcium carbonate lost per unit of nitrogen applied, a moderate rate of nitrogen (say 100 kg/ha-year) would result in almost as much additional loss as for the natural loss from rainfall alone. In any case, the total calcium loss at a 100 kg rate of nitrogen would amount to a lime ($CaCO_3$) requirement of 3,930 + 2,600 = 6,530 kg ha^{-1} of $CaCO_3$ after ten years of continuous corn.

On the conventional corn after ten years, the losses with no nitrogen were 1,350 kg $CaCO_3$/ha and the average loss per kg of nitrogen was only 0.88 kg of $CaCO_3$ so that the total $CaCO_3$ required after ten years would be 1,350 + 800 = 2,230 kg $CaCO_3$/ha. The difference due to no-tillage was 6,530 - 2,230 = 4,300 kg $CaCO_3$/ha, or 430 kg $CaCO_3$ each year. The calculations are for only one soil under climatic conditions in Kentucky, but they probably are reasonably valid for any climate with a large excess of winter and spring rainfall over evapotranspiration. That is to say, no-tillage under these conditions will acidify a soil more quickly. However, under different conditions of climate this does not seem to be the case. Data from Lal (1981) suggest that, under no-tillage, the calcium level of a west African soil tended to increase over that found in conventionally tilled soils. Similar results have been observed in southern Brazil by Muzilli (1981). Muzilli's results are shown in Table 9-9. In his cropping sequence, wheat-soybeans would be expected to cause less acidification than continuous corn (Table 9-9), since less nitrogen is used. Also, the climate of southern Brazil is not as wet during the cool season so that less overall leaching occurs. The difference in calcium in favor of no-tillage is slight but consistent for the first 10 cm of depth. This is also reflected in the pH. For the same cropping system of wheat-soybeans in Georgia, where the climate is slightly drier during the winter than in Kentucky, results of Hargrove *et al.* (1982) are shown in Table 9-10. The calcium at the surface is higher with no-tillage than with conventional tillage but is considerably lower farther down in the profile. The soil was limed twice during the four-year study so that these data are not reflective of calcium loss, per se. In fact, we (Blevins *et al.*, 1978) have similar data for two soils in Kentucky which show a higher pH at the surface with no-tillage, since the lime is simply left at the surface and moves down slowly. (The Kentucky data are shown in Table 5-14.)

As calcium decreases, this loss is partially made up by the appearance of soil acidity, principally exchangeable aluminum. At the same time, manganese becomes more available as well, but at a much lower level. The soil pH at first declines rapidly but then stabilizes near pH 4. Once this occurs, pH is a poor indicator of further damaging acidity. For example, in our study between the years of 1975 and 1980, the soil pH changed hardly at all, but the exchangeable aluminum level about doubled (Blevins *et al.*, 1977; 1983).

Exchangeable aluminum appearance plotted against exchangeable calcium loss is plotted in Figure 9-5. At first, loss of calcium has no effect on exchangeable aluminum, but after a calcium loss of approximately 2 meq/100 g, aluminum becomes exchangeable at a 1:1 ratio for each calcium lost. The cations which take the place of the

Table 9-9. Effects of four years of no-tillage and conventional tillage on chemical properties of a red latosol from Paraná, Brazil in a wheat-soybean double crop. (From Muzilli, 1981.)

DEPTH	Ca		pH		K		AVAILABLE P		Mg	
	NO-TILL	CONVEN-TIONAL	NO-TILL	CONVEN-TIONAL	NO-TILL	CONVEN-TIONAL	NO-TILL	CONVEN-TIONAL	NO-TILL	CONVEN-TIONAL
(cm)	(meq/100 g)				(meq/100 g)		(ppm)		(meq/100 g)	
0–5	6.54	5.64	6.0	5.5	0.81	0.70	49.5	12.4	2.33	1.78
5–10	5.96	5.65	5.7	5.5	0.63	0.58	22.6	10.0	2.11	1.81
10–15	5.59	5.51	5.4	5.5	0.46	0.54	12.3	9.4	1.78	1.68
15–20	5.32	5.28	5.3	5.4	0.36	0.45	7.4	7.4	1.60	1.58
20–25	4.82	4.93	5.4	5.3	0.36	0.35	7.4	5.6	1.54	1.45
25–30	4.41	4.75	–	–	0.32	0.35	5.6	5.4	1.56	1.34

Table 9-10. Effects of four years of no-tillage and conventional tillage on chemical properties of Cecil sandy loam in Georgia in a wheat-soybean double crop. (Reproduced with the permission of Hargrove *et al.*, from *Agronomy Journal* 74:684–687 [1982].)

	Ca		pH		K		P		Mg	
DEPTH	NO-TILL	CONVEN-TIONAL	NO-TILL	CONVEN-TIONAL	NO-TILL	CONVEN-TIONAL	NO-TILL	CONVEN-TIONAL	NO-TILL	CONVEN-TIONAL
(cm)	(meq/100 g)				(meq/100 g)		(ppm)		(meq/100 g)	
0–7.5	5.38	3.18	6.1	6.2	0.33	0.41	33	15	0.42	0.32
7.5–15	2.44	3.50	5.6	6.1	0.25	0.34	12	12	0.28	0.33
15–30	1.72	2.28	5.5	5.9	0.31	0.34	3	5	0.42	0.23

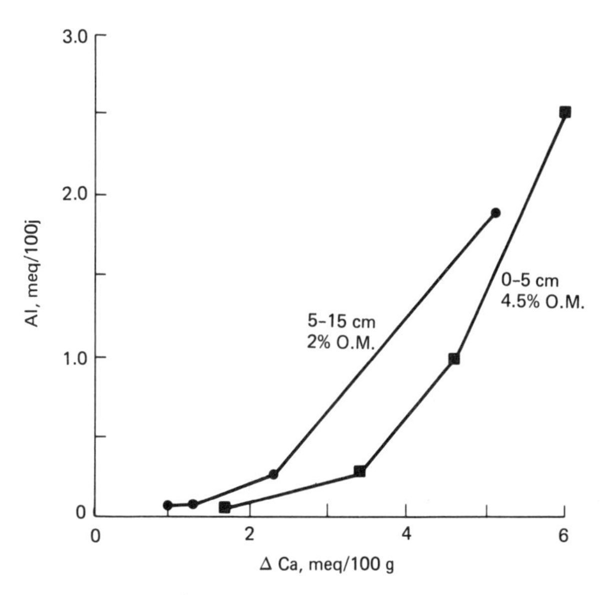

Figure 9-5. The relationship between loss of exchangeable calcium and the increase in exchangeable aluminum in Maury silt loam. (From Blevins *et al.*, 1983; Thomas, 1976)

displaced calcium are hydroxy aluminum polycations which cannot be displaced by the *N* KCl extracting solution. It also should be noted that it takes a slightly larger loss of calcium to trigger aluminum on exchange sites with the 0-5-cm layer (4.5 percent organic matter) than with the 5-15-cm layer (2.0 percent organic matter). This effect of organic matter on hydrolysis of aluminum is very important in protecting plants from aluminum toxicity (Thomas, 1975; Hargrove and Thomas, 1981).

The practical implications of these data are that if soil acidity is allowed to increase for some time there will be serious effects on yield. However, for the first few years there will be little effect on yield even if the soil pH drops significantly. In other words, the loss of calcium is a better indicator of serious acidity problems than is pH, although the drop in pH indicates that troubles are approaching.

Figure 9-6 presents exchangeable Mn in soil versus pH after ten years of no-tillage. It is obvious that there is a good relationship, but it is not a unique one because pH is variable from year to year, while manganese and aluminum keep on increasing as acidification proceeds. Although the manganese level is an order of magnitude lower than

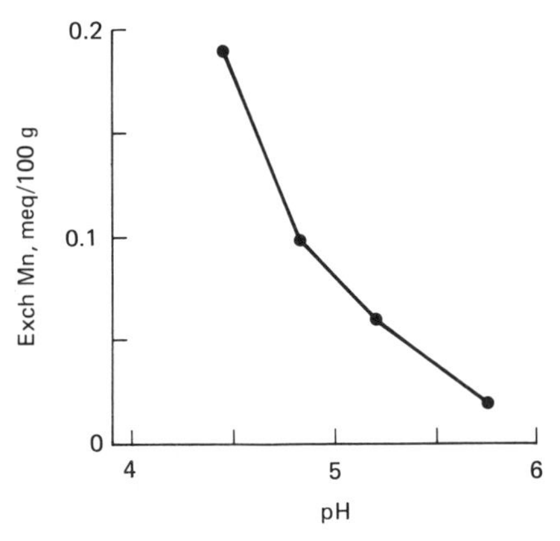

Figure 9-6. The relationship between soil pH and exchangeable manganese in Maury silt loam. (From Blevins *et al.*, 1983).

aluminum, the general tendency of manganese to cause plant injury appears to be higher than aluminum on a meq basis.

In summary, loss of calcium, appearance of exchangeable aluminum and manganese and decrease in pH are closely related. Changes in pH occur quickly, but exchangeable aluminum and manganese continue to increase even after pH no longer changes. In the humid zone of the U.S., the more rapid acidification of soils under no-tillage probably is the most notable chemical property which is seen in the field. Unfortunately it is a disadvantage for no-tillage.

In areas where there is not a large excess of rain over evapotranspiration, this effect of no-tillage on acidity is reversed. For example, Lal (1981) found that average pH for continuous corn (two crops per year) using no-tillage was 5.3 versus 5.0 for plowed soils. This also is seen in Brazil (Table 9-9), as discussed earlier. It is obvious that much work remains to be done to relate the soil acidity to soil characteristics, to climate and to cropping system used. There seem to be strong interactions between all these factors but the data are so sparse that guessing is about the best that can be done at present.

Phosphorus and Potassium

In terms of the concentration of the nutrients phosphorus and potassium with depth, the results come much closer to being consistent. The reason for the results appears to be the fact that as a rule both phosphorus and potassium are applied to the soil surface. Then, with conventional tillage, they are inverted or partly inverted and then mixed further. This mixing goes on year after year until both phosphorus and potassium are pretty evenly mixed to the depth of tillage. Notice the results from both Brazil and Georgia (Tables 9-9 and 9-10), which show a slow decline in concentrations of phosphorus and potassium with depth under conventional tillage. Table 9-11 also has potassium data for Kentucky.

On the other hand, with no-tillage, the surface-applied fertilizer is not moved except by water and (much more slowly) by diffusion. Hence, the concentrations of phosphorus and potassium are highest by far in the upper 5 cm of soil and then decline rapidly with soil depth. It is most common for there to be no real difference in total exchangeable potassium in the surface 30 cm, but to be distributed differently. With phosphorus, however, surface application without further mixing gives much higher totals over the 0–30-cm depth. This apparently occurs because, in the absence of mixing, there is a "banding" effect of surface application; or, in other words, there is less phosphorus fixation. Notice that data from both Brazil and Georgia (Tables 9-9 and 9-10) show much higher totals for phosphorus under no-tillage than with conventional tillage. As discussed in Chapter 5, phosphorus uptake is also higher with no-tillage, suggesting that the difference in total soil test phosphorus is a real one.

Magnesium

There are small and non-consistent differences in exchangeable magnesium between no-tillage and conventional tillage. Data in Tables 9-8 and 9-11 show that in Kentucky we have found essentially no difference, which is surprising considering the large differences in calcium found. In Georgia, where dolomitic lime was applied during the experiment, there was a slightly higher magnesium level under no-tillage. In Brazil, there was a rather large difference in favor of no-tillage in the surface 5 cm, but the difference was reduced rather quickly with depth.

It appears from these limited data that the difference in magnesium are too small to make a difference practically and that, in any case, there is no detrimental effect of no-tillage on soil magnesium.

Table 9-11. Effects of ten years of no-tillage and conventional tillage with corn grown continuously on chemical properties of Maury silt loam in Kentucky. (From Blevins *et al.*, 1983.)

	Ca		pH		K		Mg	
DEPTH	NO-TILL	CONVENTIONAL	NO-TILL	CONVENTIONAL	NO-TILL	CONVENTIONAL	NO-TILL	CONVENTIONAL
(cm)	(meq/100 g)					(meq/100 g)		
0–5	3.85	7.30	5.20	6.40	0.59	0.33	0.68	0.67
5–15	5.66	7.06	5.90	6.35	0.39	0.36	0.64	0.64
15–30	6.39	7.39	6.28	6.45	0.20	0.26	0.55	0.65

Summary of Chemical Properties

It is difficult to summarize all the chemical properties in soil as they are affected by no-tillage. For example, a loss of calcium, a decrease in pH and an increase in exchangeable aluminum and manganese are always found in the humid region of the U.S. This propensity is hastened by increased rates of nitrogen. In other areas where there is no great excess of rain over evapotranspiration, the loss of calcium is not seen and, in fact, there is a slight increase in calcium and pH with no-tillage. It is clear that much more needs to be done to determine the causes for these differences. In the meantime, it can be reasoned that where there is enough water during the year to remove $Ca(NO_3)_2$, no-tillage will lose more of this salt than conventional tillage.

As far as the distribution of phosphorus and potassium is concerned, the data are much more straightforward. In most cases, these nutrients are applied to the soil surface and are not moved by soil manipulation. Therefore, their concentrations are higher at the soil surface than they are with conventional tillage where they are mixed. In the case of potassium, the total amount found in the surface 30 cm of soil is about the same as the total found with conventional tillage, only the distribution is different. With phosphorus, however, the distribution is towards the surface with no-tillage and the total available phosphorus is also higher, a result of less mixing (and fixation). Finally, soil magnesium is not affected appreciably by tillage. It may be slightly higher with no-tillage or about the same, but the differences are not significant as far as influencing plant growth.

REFERENCES

Balderston, W. L., B. Sherr, and W. J. Payne. 1976. Blockage by acetylene of nitrous oxide reduction in *Pseudomonas perfectomarinus*. *Appl. Environ. Microbiol.* **31**:504–508.

Barber, S. A. 1979. Corn residue management and soil organic matter. *Agron. J.* **71**:625–627.

Barber, D. A. and C. J. Standell. 1977. Preliminary observations on the effects of direct drilling on the microbial activity of soil. *Agric. Res. Coun. (G.B.) Letcombe Lab. Annu. Rep.* **1976**:58–60.

Barnes, B. T. and F. B. Ellis. 1979. Effects of different methods of cultivation and direct drilling, and disposal of straw residues, on populations of earthworms. *J. Soil Sci.* **30**:669–679.

Bennett, O. L., G. Stanford, E. L. Mathias, and P. E. Lundberg. 1975. Nitrogen conservation under corn planted in quackgrass sod. *J. Environ. Qual.* **4**:107–110.

Bertsch, P. M. 1982. Unpublished data. Kentucky Agric. Exp. Sta., Lexington, Kentucky.

Blevins, R. L., L. W. Murdock, and G. W. Thomas. 1978. Effect of lime application on no-tillage and conventionally tilled corn. *Agron. J.* **70**:322–326.

Blevins, R. L., G. W. Thomas, and P. L. Cornelius. 1977. Influence of no-tillage and nitrogen fertilization on certain soil properties after five years of continuous corn. *Agron. J.* **69**:383–386.

Blevins, R. L., G. W. Thomas, M. S. Smith, W. W. Frye, and P. L. Cornelius. 1983. Changes in soil properties after 10 years non-tilled and conventionally tilled corn. *Soil and Tillage Res.* **3**:135–146.

Bremner, J. M. and A. M. Blackmer. 1979. Effects of acetylene and soil water content on emission of nitrous oxide from soils. *Nature* **280**:380–381.

Brown, P. L. and D. D. Dickey. 1970. Losses of wheat straw residue under simulated field conditions. *Soil Sci. Soc. Am. Proc.* **34**:118–121.

Burford, J. R., R. J. Dowdell, and R. Crees. 1976. Mineralization of nitrogen in a clay soil (Denchworth series). *Agric. Res. Coun. (G.B.) Letcombe Lab. Annu. Rep.* **1976**:84.

Burford, J. R., R. J. Dowdell, and R. Crees. 1977. Denitrification: Effect of cultivation on fluxes of nitrous oxide from the soil surface. *Agric. Res. Coun. (G.B.) Letcombe Lab. Annu. Rep.* **1977**:71–72.

Burford, J. R., R. J. Dowdell, and R. Crees. 1981. Emission of nitrous oxide to the atmosphere from direct-drilled and ploughed clay soils. *J. Sci. Food Agric.* **32**:219–223.

Cannell, R. Q. 1981. Potentials and problems of simplified cultivation and conservation tillage. *Outlook on Agric.* **10**:379–384.

Cannell, R. Q. and F. B. Ellis. 1977. Review of progress on research on reduced cultivation and direct drilling. *Agric. Res. Coun. Letcombe Lab. Annu. Rep.* **1976**:25–27.

Cannell, R. Q., F. B. Ellis, D. G. Christian, J. P. Graham, and J. T. Douglas. 1980. The growth and yield of winter cereals after direct drilling, shallow excavation and ploughing on non-calcareous clay soils, 1974–78. *J. Agric. Sci.* (Cambridge) **94**:345–359.

Cochran, V. L., L. F. Elliott, and R. I. Papendick. 1977. The production of phytotoxins from surface crop residues. *Soil Sci. Soc. Am. J.* **41**:903–908.

Cochran, V. L., L. F. Elliott, and R. I. Papendick. 1980. Carbon and nitrogen movement from surface-applied wheat (*tritium aestivum*) straw. *Soil Sci. Soc. Am. J.* **44**:978–982.

Corbett, D. C. M. and R. M. Webb. 1970. Plant and soil nematode population changes in wheat grown continuously in ploughed and in unploughed soil. *Ann. Appl. Biol.* **65**:327–335.

Dancer, W. S., L. A. Peterson, and G. Chesters. 1973. Ammonification and nitrification of N as influenced by soil pH and previous N treatments. *Soil Sci. Soc. Am. Proc.* **37**:67–69.

Darwin, Charles. 1907. *The formation of vegetable mould through the action of worms.* D. Appleton and Company, New York, N.Y.

Dawson, R. C. 1945. Effect of crop residues on soil micropopulations, aggregation, and fertility under Maryland conditions. *Soil Sci. Soc. Am. Proc.* **10**: 180–184.

Dawson, R. C., V. T. Dawson, and T. M. McCalla. 1948. Distribution of microorganisms in the soil as affected by plowing and subtilling crop residues. *Res. Bull. 155,* University of Nebraska, Lincoln.

Doran, J. W. 1980a. Microbial changes associated with residue management with reduced tillage. *Soil Sci. Soc. Am. J.* **44**:518–524.

Doran, J. W. 1980b. Soil microbial and biochemical changes associated with reduced tillage. *Soil Sci. Soc. Am. J.* **44**:765–771.

Douglas, C. L., Jr., R. R. Allmaras, P. E. Rasmussen, R. E. Ramig, and N. C. Roager, Jr. 1980. Wheat straw composition and placement effects on decomposition in dryland agriculture of the Pacific Northwest. *Soil Sci. Soc. Am. J.* **44**:833–837.

Douglas, J. T. and M. J. Goss. 1982. Stability and organic matter content of surface soil aggregates under different methods of cultivation and in grassland. *Soil and Tillage Res.* **2**:155–175.

Douglas, J. T., M. J. Goss, and D. Hill. 1980. Measurements of pore characteristics in a clay soil under ploughing and direct drilling including use of a radioactive tracer (^{144}Ce) technique. *Soil and Tillage Res.* **1**:11–18.

Dowdell, R. J. and R. Q. Cannell. 1975. Effect of plowing and direct drilling on soil nitrate contents. *J. Soil Sci.* **26**:53–61.

Dowdell, R. J. and R. Crees. 1980. The uptake of 15N-labelled fertilizer by winter wheat and its immobilization in a clay soil after direct drilling or ploughing. *J. Sci. Food and Agric.* **31**:992–996.

Ehlers, W. 1975. Observations on earthworm channels and infiltration on tilled and untilled loess soil. *Soil Sci.* **119**:242–249.

Ehlers, W. 1979. Influence of tillage on hydraulic properties of loessial soils in western Germany. In *Soil Tillage and Crop Production.* R. Lal, Editor. Proceedings series No. 2, IITA, Ibadan, Nigeria, pp. 33–45.

Elliott, L. F., V. L. Cochran, and R. I. Papendick. 1981. Wheat residue and nitrogen placement effect on wheat growth in the greenhouse. *Soil Sci.* **131**: 48–52.

Ellis, F. B., J. G. Elliott, F. Pollard, R. Q. Cannell, and B. T. Barnes. 1979. Comparison of direct drilling, reduced cultivation and ploughing on the growth of cereals. 3. Winter wheat and spring barley on a calcareous clay. *J. Agric. Sci.* (Cambridge) **93**:391–401.

Fairchild, D. M. and T. E. Staley. 1979. Tillage method effects on soil microbiota and C/N reservoirs. *Abstracts Am. Soc. Agron., 1979.* Fort Collins, Colorado, P. 157.

Focht, D. D. 1978. Methods for analysis of denitrification in soils. In *Nitrogen in the Environment.* D. R. Nielsen and J. G. McDonald, Editors. Academic Press, New York.

Gantzer, C. J. and G. R. Blake. 1978. Physical characteristics of Le Sueur clay loam soil following no-till and conventional tillage. *Agron. J.* **70**:853–857.
Giddens, J. 1957. Rate of loss of carbon from Georgia soils. *Soil Sci. Soc. Am. Proc.* **21**:513–515.
Greilich, D. and E. M. Klimanek. 1976. Zum einflub unterschiedlicher intensitat der bodenbearbeitung auf den O_2 und CO_2 gehalt der bondenluft sowoe auf einige bodenbiologische kenwerete. *Arch. Ackeru. Pflenzenbau u. Bodenkd.* **20**(*3*):177–186.
Hargrove, W. L. and G. W. Thomas. 1981. Effect of organic matter on exchangeable aluminum and plant growth in acid soils. Chapter 8 in *Chemistry in the Soil Environment. Am. Soc. Agron.,* Special Publication No. 40, Madison, Wisconsin, pp. 151–166.
Hargrove, W. L., J. T. Reid, J. T. Touchton, and R. N. Gallaher. 1982. Influence of tillage practices on the fertility status of an acid soil double-cropped to wheat and soybeans. *Agron. J.* **74**:684–687.
Hice, T. H. and R. L. Todd. 1980. Decomposition of corn residue under conventional and minimum tillage practices. *Abstracts Am. Soc. Agron.* Detroit, Michigan, p. 154.
Hoyt, G. D., R. L. Todd, and R. A. Leonard. 1980. Nitrogen cycling in conventional and no-tillage systems in the southeastern Coastal Plains. *Abstracts Am. Soc. Agron.* Detroit, Michigan, p. 154.
Kevan, D. K. M. 1968. *Soil Animals.* H. F. and G. Witherby, London, p. 162.
Kladivko, E. and D. Keeney. 1981. Nitrogen transformations under conservation tillage systems. *Proc. of the 1981 Fertilizer, Aglime and Pest Management Conference,* Madison, Wisconsin.
Klein, T. M. and J. S. Koths. 1980. Urease, protease, and acid phosphatase in soil continuously cropped to corn by conventional or no-tillage methods. *Soil Biol. Biochem.* **12**:293–294.
Kruckelmann, H. W. 1975. Effects of fertilizers, soils, soil tillage, and plant species on the frequency of Endogone chlamydospores and mycorrhizal infection in arable soils. In *Endomycorrhizas,* F. E. Sanders, B. Mosse, and P. B. Tinker, Editors. Academic Press, New York.
Lal, R. 1976. No-tillage effects on soil properties under different crops in Western Nigeria. *Soil Sci. Soc. Am. Proc.* **40**:762–768.
Lal, R. 1981. No tillage farming in the tropics. In *No-Tillage Research: Research Reports and Reviews.* R. E. Phillips, G. W. Thomas and R. L. Blevins, Editors University of Kentucky, pp. 103–151.
Lal, R., D. De Vleeschauwer and R. M. Nganje. 1980. Changes in properties of a newly cleared tropical Alfisol as affected by mulching. *Soil Sci. Soc. Am. J.* **44**:827–833.
Lucas, R. E., J. B. Holtman, and L. J. Conner. 1977. Soil carbon dynamics and cropping systems. In *Agriculture and Energy.* William Lockeretz Editor. Academic Press, New York, pp. 333–351.
Lynch, J. M. and L. M. Panting. 1980a. Cultivation and the soil biomass. *Soil Biol. Biochem.* **12**:29–33.

Lynch, J. M. and L. M. Panting. 1980b. Variation in the size of the soil biomass. *Soil Biol. Biochem.* **12**:547–550.

Mannering, J. V., D. R. Griffith, and C. B. Richey. 1975. Tillage for moisture conservation. Paper No. 75-2523. *Am. Soc. Agric. Eng.*, St. Joseph, Michigan.

McCalla, T. M. 1958. Microbial and related studies of stubble mulching. *J. Soil Water Conser.* **13**:255–258.

Mengel, D. B. 1980. Nitrogen management in reduced tillage systems. *Indiana Plant Food and Agricultural Chemicals Conference.* West Lafayette, Indiana.

Mitchell, D. J., N. C. Schenck, D. W. Dickson, and R. N. Gallaher. 1980. The influence of minimum tillage on populations of soilborne fungi, endomycorrhizal fungi, and nematodes in oats and vetch. In *Proceedings of the Third Annual No-Tillage Systems Conference,* R. N. Gallaher, Editor, Gainesville, Florida, pp. 115–119.

Muzzilli, O. 1981. Manejo da fertilidade do solo. In *Fundacão Instituto Agronomico de Paraná. Plantio directo no estado do Paraná Circular IAPAR* **23**, 244 pp.

Norstadt, F. A. and T. M. McCalla. 1968. Microbially induced phytotoxicity in stubble-mulched soil. *Soil Sci. Soc. Am. Proc.* **32**:241–245.

Norstadt, F. A. and T. M. McCalla. 1969. Microbial populations in stubble-mulched soil. *Soil Sci.* **107**:188–193.

Olson, G. W. and T. M. McCalla. 1960. A comparison of microbial respiration in soils after a 20-year period of subtilling or plowing. *Soil Sci. Soc. Am. Proc.* **24**:349–352.

Parker, D. T. 1962. Decomposition in the field of buried and surface applied cornstalk residue. *Soil Sci. Soc. Am. Proc.* **26**:559–562.

Parker, D. T., W. T. Larson, and W. V. Bartholomew. 1957. Studies on nitrogen tie-up as influenced by location of plant residues in soils. *Soil Sci. Soc. Am. Proc.* **21**:608–612.

Pidgeon, J. D. and B. D. Soane. 1977. Effects of tillage and direct drilling on soil properties during the growing season in a long-term barley mono-culture system. *J. Agric. Sci.* (Cambridge) **88**:431–442.

Powlson, D. S. 1980. Effect of cultivation on the mineralization of nitrogen in soil. *Plant and Soil* **57**:151–153.

Rice, C. W. and M. S. Smith. 1982a. Unpublished data. Kentucky Agric. Exp. Sta., Lexington, Kentucky.

Rice, C. W. and M. S. Smith. 1982b. Denitrification in plowed and no-till soils. *Soil Sci. Soc. Am. J.* **46**:1168–1173.

Rovira, A. D. and E. L. Greacen. 1957. The effect of aggregate disruption on the activity of micro-organisms in the soil. *Aust. J. Agric. Res.* **8**:659–673.

Sain, P. and F. E. Broadbent. 1977. Decomposition of rice straw in soils as affected by some management factors. *J. Environ. Qual.* **6**:96–100.

Shear, G. M. and W. W. Moschler. 1969. Continuous corn by no-tillage and conventional tillage methods: A six-year comparison. *Agron. J.* **61**:524–526.

Smith, M. S. 1979. Unpublished data. Kentucky Agric. Exp. Sta., Lexington, Kentucky.

Smith, M. S., M. K. Firestone, and J. M. Tiedje. 1978. The acetylene inhibition method for short-term measurement of soil denitrification and its evaluation using nitrogen-13. *Soil Sci. Soc. Am. J.* **42**:611–615.

Smith, M. S. and K. Zimmerman. 1981. Nitrous oxide production by non-denitrifying soil NO_3^- reducers. *Soil Sci. Soc. Am. J.* **45**:865–871.

Staley, T. E. and D. M. Fairchild. 1979. Enumeration of denitrifiers in an Appalachian soil. *Abstracts Annual Meeting of Am. Soc. for Microbiol.,* Los Angeles.

Thomas, G. W. 1975. The relationship between organic matter content and exchangeable aluminum in acid soil. *Soil Sci. Soc. Am. Proc.* **39**:591.

Thomas, G. W. 1976. Unpublished data. Kentucky Agric. Exp. Sta., Lexington, Kentucky.

Thomas, G. W., R. L. Blevins, R. E. Phillips, and M. A. McMahon. 1973. Effect of a killed sod mulch on nitrate movement and corn yield. *Agron. J.* **65**: 736–739.

Thomas, S. H. 1978. Population densities of nematodes under seven tillage regimes. *J. Nematol.* **10**:24–27.

Triplett, G. B., Jr., F. Haghira, and D. M. Van Doren, Jr. 1979. Plowing effect on corn yield response to N following alfalfa. *Agron. J.* **71**:801–803.

Triplett, G. B., Jr., D. M. Van Doren, Jr., and B. L. Schmidt. 1968. Effect of corn (*Zea mays* L.) stover mulch on no-tillage corn yield and water infiltration. *Agron. J.* **60**:236–239.

Tyler, D. D. and G. W. Thomas. 1977. Lysimeter measurements of nitrate and chloride losses from soil under conventional and no-tillage corn. *J. Environ. Qual.* **6**:63–66.

Van Ouwerkerk, C. and F. R. Boone. 1970. Soil-physical aspects of zero-tillage experiments. *Neth. J. Agric. Sci.* **18**:247–261.

Vieira, M. J. 1981. Plantio directo no estado do Paraná. In *Circular IAPAR No. 23.* Londrina – Paraná, Fundacao Instituto Agronomico Do Paraná, pp. 19–32.

Wacha, A. G. and K. H. Tiffany. 1979. Soil fungi isolated from fields under different tillages and weed-control regimes. *Mycologia* **71**:1215–1226.

Wallwork, J. A. 1976. *The Distribution and Diversity of Soil Fauna.* Academic Press, New York.

Yoshinari, T. and R. Knowles. 1976. Acetylene inhibition of nitrous oxide reduction by denitrifying bacteria. *Biochem. Biophys. Res. Comm.* **69**:705–710.

10
Multicropping

Shirley H. Phillips
Associate Director of Extension
College of Agriculture, University of Kentucky

and

Grant W. Thomas
Professor of Agronomy
University of Kentucky

MULTICROPPING WITH NO-TILLAGE TECHNIQUES

The potential for multicropping may be the most important factor in no-tillage agriculture. The advantages of the no-tillage system became more important with multicropping, and these include: (1) reduced labor and costs; (2) elimination of moisture loss associated with conventional tillage at planting time, ensuring stands of second and third crops under restricted rainfall patterns; (3) production of more than one crop per year, which increases land use; (4) further reduction of soil erosion; (5) maintenance of soil structure by elimination of plowing and land preparation; (6) time saved in planting the second and third crops when timeliness of planting is very important; (7) improved cash flow by producing additional crops per year; (8) increased use of equipment, therefore decreasing cost investment per hectare; and (9) interseeding of legumes into grass pastures with minimum reduction of grass stands.

With the no-tillage system, harvesting can be followed immediately by planting of the succeeding crop, thus reducing the time lag between crops. On well-drained soils, no-tillage crops can be planted over a wider range of soil moisture conditions than can conventional tillage crops, and this further reduces the time before the next crop can be planted in the multicropping sequence.

Multiple cropping is considered to be any scheme which results in more than a single crop being harvested from a piece of land during one year. Thus, it includes both intercropping and sequential cropping. With few exceptions, intercropping is not extensively practiced in the United States. It is the most important form of multiple cropping in tropical countries, however. Sequential multiple cropping with short-term vegetable crops has been important in parts of the United States with long growing seasons for many years. A major increase in multiple cropping in the U.S. has occurred in the last 15 years and traces its success to several factors. These include plant breeding, farm mechanization, irrigation (in places), herbicides, high land cost and relatively poor prices for farm grains and introduction of no-tillage systems.

The potential for multiple cropping, however, has barely been touched. Although it does have a northern limit of adoption in the U.S., that limit continues in retreat. Or, to put it another way, no one really knows what the limit is. Many systems have been tried and a few are commercially viable and even common. Part of the challenge for the future of multiple cropping is to devise more profitable systems for various parts of the country. These systems need to be reasonably dependable, more profitable than monocropping and not so intricate that people with common sense cannot succeed in mastering them.

Substitution of capital and labor for land is a valid reason for the rapid growth in multicropping. Limited land resources or higher land values cause additional hectares to be multicropped each year. In a double cropping system such as small grain-soybeans, 300 ha become 600 ha, or in the tropical areas with small grain-soybeans-grain sorghum triple cropped, 300 ha enlarges to a potential 900 ha of crops.

Traditional zones for multicropping can be moved by the adoption of no-tillage. The wheat-soybean double cropping system was primarily used in the southern states of the U.S. prior to wide adoption of no-tillage and is now common to Ohio, Illinois, Indiana and Iowa.

Multicropping must be geared to climatic conditions and will vary according to temperature, rainfall and other conditions in the various areas of world production.

Innovations and practices will improve chances of success such as early harvest by swathing, drying of grain, earlier varieties geared to multicropping with increased demand for food and fiber.

The following conclusions can be made after comparing swathing and combining to direct combining of wheat at two climatically diverse locations in Kentucky in 1974 and 1975 (TeKrony *et al.*, 1976).

1. When the wheat was swathed at a moisture content of 40–42 percent or less, there was no reduction in yield, test weight or percent germination of the seed compared to the direct combined check.
2. By swathing at 40–42 percent moisture content compared to direct combining, the harvest date of the wheat was advanced 7–10 days in 1974 and 4 days in 1975.
3. By swathing the wheat at 42 percent moisture content in 1974, the number of wild garlic bulblets present in the combine-run seed was reduced 40–50 percent compared to the check.

It appears that this practice may be feasible to Kentucky farmers as a procedure for expediting wheat harvest. This will allow earlier planting of soybeans following wheat harvest in a double crop situation which will mean increased soybean yields. Swathing may be especially appealing to farmers who double crop a large number of hectares, since it will allow them to spread out the time of wheat harvest and soybean planting.

Multicropping is not without risk. Herbicide selection becomes more complex by increasing the weed species grown on the same land. Disease incidence may increase with an annual production of the same species on the same field each year. Higher levels of capital and management are required and timing becomes paramount for successful operations.

Many schemes are available to combat the time element. The most popular multicropping system in the U.S. involves small grain grown in the winter months followed by soybeans in the summer. Growers can use barley and wheat to spread harvest dates and use varieties with different maturity ranges within the species to further spread harvest of the small grain and planting of the soybeans.

Aerial seeding of small grains into standing soybeans near leaf drop is used to gain two to three weeks in establishing the small grain following the soybean crop.

In the U.S. grain farmers follow these major systems:

Two crops/year – double cropping – winter small grain-soybeans
– winter small grain-grain sorghum
Three crops in two years – corn-small grain-soybeans.

Livestock farms use small grain for winter grazing, harvest the same small grain for silage in the early spring and follow with corn for ensiling. Yields are increased by 3–5 mt/ha while providing a summer silage feed during the dormant period for most cool-season grasses and legumes.

Harvest of the first hay crop followed by corn as an interim crop can be used as stands of red clover, alfalfa, other legumes or grasses become depleted prior to reestablishment.

BEGINNINGS OF MULTIPLE CROPPING IN THE U.S.

The origins of multiple cropping in the U.S. are directly traceable to the Indians, who taught the first settlers their use of corn as a support crop for beans and squash. Intercropping of these species was very common in this country until quite recently. However, there is absolutely no indication that a plot of land so planted is more productive in terms of edible dry matter produced than the same crops arranged separately. In fact, it is likely that it generally is less productive.

As commercial vegetable production moved to climatic areas where successive cropping was practical, multiple cropping became a normal way to use land. Because irrigation development is expensive in such states as Florida and California, sequential cropping makes maximum use of the expensive irrigated land.

The movement towards multiple cropping in extensive field crops was influenced by several factors including mechanization, herbicides, high farm-land prices and continuing low prices received for grain crops.

The most popular multiple cropping sequence in the U.S. today, whether measured in land area in crops, profitability or dependability, is winter wheat followed by soybeans. This system depends on modern planting, harvesting, handling and storage equipment that requires few modifications in going from wheat to corn to soybeans. Once this equipment is in place, there are only two ways to make maximum returns with it. One is to crop more hectares, and the other is to have the same hectares grow more crops. In many cases, both routes are taken.

In addition to equipment which assures timely planting and harvesting and allows drying of wet grains and short-term storage, multiple grain cropping is dependent on varieties that mature at predictable early times and herbicides which will not prevent growth of the following crop. Intimately connected with multiple cropping systems such as wheat-soybeans are systems of reduced tillage. Although the use of reduced tillage may be motivated by energy savings and erosion prevention in some instances, the saving of time between crops is the most important role in multiple cropping. A secondary factor is reduction in soil water loss when limited or no-tillage is practiced in establishing the second crop.

Other cropping sequences which have the potential of adding millions of hectares of land to multiple cropping are wheat-corn, wheat-sorghum, wheat-cotton, corn-corn, corn-soybeans and corn-sorghum. Other possibilities are included in Table 10-1.

REQUIREMENTS FOR SUCCESSFUL MULTIPLE CROPPING

Length of Growing Season

It is obvious that multiple cropping for grain depends on a reasonably long frost-free season, one that averages 180 days or more. In the typical system of wheat-soybeans, with wheat harvested somewhere around the last week in June, the longest days of the year are lost almost entirely to dry matter production because, even with narrow rows, a full canopy of soybeans will not be established for about six weeks (see Table 2-5 for net solar radiation by month), sometime after the first of August. The growth of corn during that six-week period would have been at a maximum, whereas almost no dry matter

Table 10-1. Some multiple cropping possibilities.

Corn – Corn
Corn – Grain Sorghum
Grain Sorghum – Ratoon – Grain Sorghum
Wheat – Soybeans
Wheat – Corn
Wheat – Millet
Wheat – Grain Sorghum
Wheat – Grain Sorghum – Soybeans
Barley – Soybeans
Barley – Corn
Wheat (silage) – Corn (silage)
Winter Legume (for N) – Corn
Wheat – Red Clover
Grass – Clover
Grass – Alfalfa
Wheat – Cotton
Wheat – Sugar Beets
Vegetable – Vegetable – Vegetable
Sugar Beets – Sorghum
Rice – Ratoon – Rice
Green Pea – Sweet Corn
Annual Rye Grass (grazed and allowed to reseed) – Soybeans
Winter Small Grain – interseeded into cool season grass and legume pasture
Winter Small Grain – interseeded into summer perennials
Summer Annuals, Forages – interseeded into cool season grasses

(and no usable product) was produced by the soybeans. The principle is illustrated well by Figure 10-1 showing that a single long-term crop out-produced three short-term crops.

Data for a Florida multiple cropping scheme (corn-soybeans-rye) as compared to sugar cane (Allen *et al.*, 1976) are shown in Table 10-2. These results illustrate the danger of supposing that multiple crops automatically are more productive than single crops. Another example is that of wheat. In Europe, where yields of wheat exceeding 6–7 mt/ha are commonplace, an often overlooked fact is that the wheat does not mature until August, thereby taking advantage of the long days (in Europe, very long days) of July.

Thus, in general, it would be advantageous to grow a single long-term crop than to grow two or three short-term crops in terms of yield. Unfortunately, in much of the middle and southern latitudes of the U.S., maturity of crops is hastened by high temperatures and

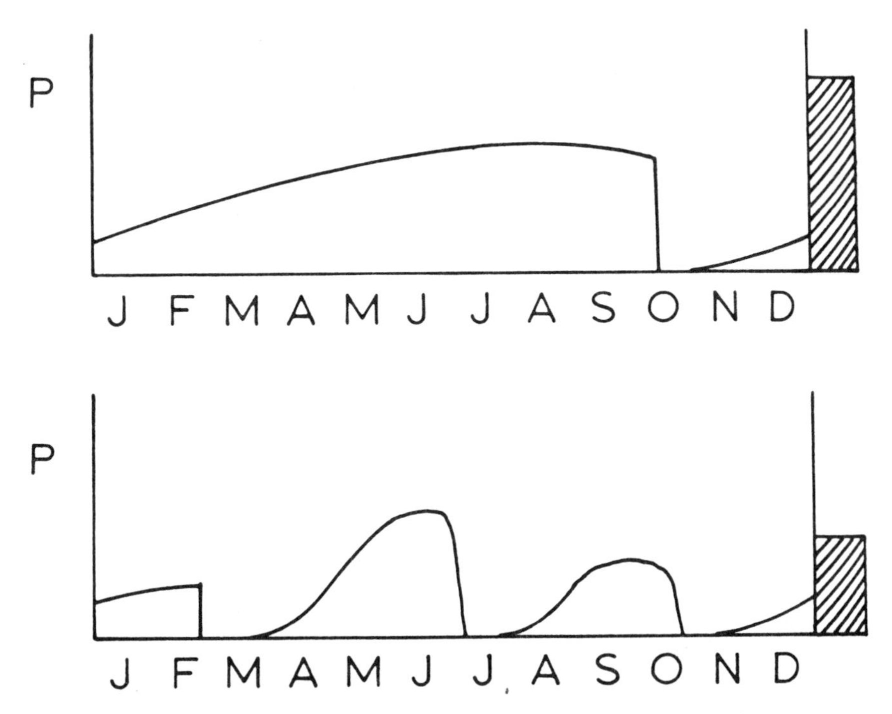

Figure 10-1. Illustration of theoretical photosynthesis or productivity, *P*, of a single, long-term crop as compared with three short-term sequential crops. The height of the bar graph at the end illustrates the relative total annual production of the area under the curves (reproduced with the permission of Allen *et al.*, from *Multiple Cropping*, Special Publication No. 27 of the American Society of Agronomy.

Table 10-2. Productivity of sugar cane, corn, soybeans and rye. (Reproduced with the permission of Allen *et al.*, from *Multiple Cropping*, Special Publication No. 27 of the American Society of Agronomy [1976].)

	LEAF PHOTOSYNTHESIS RATE	CROP GROWTH RATE	YIELD OF HARVESTABLE PRODUCT
	($mg\ CO_2/dm^2/hr$)	($g/cm^2/day$)	(mt/ha)
Sugar cane	60	25	14.0 sugar
Corn	60	25	6.0 grain
Soybeans	45	15	2.5 beans
Winter rye	30	5	0 grain

there is no suitable long-season crop such as sugar cane which will produce competitively. Hence, the combination of a cool-season and a warm-season crop, the latter delayed as little as possible.

In the most northerly extension of double cropped wheat-soybeans in the U.S. (about 40° latitude), there is a desperate haste to plant soybeans as quickly as possible to avoid losses of soybeans due to frost and to avoid yield losses from lost solar radiation. These yield losses approach 25–30 kg/day from the last week of June onward. The tactics used include swathing of wheat to promote drying of the grain and foliage and even "relay" planting of soybeans before wheat is harvested. The possibility of "relay" planting of soybeans in wheat appears to have some promise if the damage to wheat can be minimized and there is good weed control (Table 10-3).

Another tactic is to switch to barley, which matures earlier but which is less profitable than wheat and which, in humid regions, is more susceptible to disease than wheat.

Woven through all these schemes, however, is a basic understanding that soybean yield must be damaged as little as possible. Therefore, damage to wheat by relay planting, loss in yield of wheat by early harvest or even changing to a less profitable crop are accepted in place of significant soybean losses. Generally, this principle holds true in any multiple cropping system. That is, one crop will receive preferential treatment at the risk of damaging another less profitable crop.

Table 10-3. Wheat and soybean yields in no-tillage double cropped and relay planted soybeans at Lexington, Kentucky. (From Slack *et al.*, 1980.)

YEAR	METHOD OF PLANTING	WHEAT YIELD	VALUE	SOYBEAN YIELD	VALUE	TOTAL VALUE
		(mt/ha)	($/ha)	(mt/ha)	($/ha)	($/ha)
1979	Soybeans after wheat	3.4	454	2.4	565	1019
	Soybeans relay planted in wheat	1.5	200	3.6	856	1056
1980	Soybeans after wheat	3.8	499	2.0	468	967
	Soybeans relay planted in wheat	2.3	309	1.4	339	648

In a more southerly climate, say Tifton, Georgia, using the multiple cropping system corn-corn, Widstrom and Young (1980) concluded that the second corn crop would have to be utilized as silage or other forage in order to get reliable yields from it because grain yields were erratic (Table 10-4). Similarly, in Kentucky, a wheat-corn double crop would have to utilize one or both crops as silage for maximum utilization. This can meet the needs of a dairy farm but is not useful at all on a cash grain farm.

Water

In much of the region where multiple cropping is now practiced most heavily, summertime rainfall is inadequate in most years, for optimum crop yields. Changing from a single crop to a double crop does not improve the situation but it can usually be managed without total crop failure. In the most common system of wheat-soybeans, very little water is lost during the ripening period of the wheat, and this becomes a time for storage of soil water if rains occur. However, if conventional tillage practices of plowing and disking are carried out, much of the stored water will be lost by the time the soybeans are planted. This loss is due both to delay and to soil manipulation and is frequently fatal to the succeeding soybean crop due to slow germination and/or soil crusting. A system that conserves water and speeds planting is to plant directly in the wheat stubble with no soil preparation other than a slit where the seeds are placed. This not only saves water but, with the cover of wheat stubble, prevents surface crusting. In some years, with timely rain, there is

Table 10-4. Corn yields of corn-corn double cropping in south Georgia. (Reproduced with the permission of Widstrom and Young, from *Agronomy Journal* 72:302-305 [1980].)

	CORN YIELDS (mt/ha)		
	1976	1977	AVERAGE
First corn crop	8.7	7.2	8.0
Second corn crop	1.6	1.6	1.6
Total production	10.3	8.9	9.6

little effect of tillage or lack of tillage on soil water, but in other years the no-tillage system with small water savings is enough to make the difference between a crop and no crop of soybeans. This same principle applies with other cropping systems so that a mulch-no-tillage system results in less soil water loss.

In drier climates where irrigation is practiced, multiple cropping typically demands both more and more timely irrigation. This demand can very easily make multiple cropping uneconomical in semi-humid regions where cost of water delivery to fields is high. At the same time, in areas such as the high plains, where most rainfall occurs in the summer and year-to-year variation is great, a wheat-sorghum sequence could be profitable if there is high soil water at wheat harvest time. If, on the other hand, the season is so dry that water costs would make the second crop marginal, the second crop could either be delayed until sufficient rainfall occurred or simply skipped for that year. This type of decision is much easier to make if no-tillage is practiced. In addition, whatever rainfall occurs will be better saved by the wheat stubble mulch and less erosion will occur, whatever the final decision is on the use of the land for that season. Regardless of the climatic conditions and cropping system used, there is a built-in advantage for no-tillage or very reduced tillage. Unger and Stewart (1976) have pointed out that in multiple cropping systems, no-tillage saves time, water and soil (from erosion), and reduces soil crusting.

One difficulty in sustained multiple cropping is to establish a small grain in soybeans for the next year's crop. Because late-planted, full-season soybeans are growing actively until the first part of October in the south and lower midwest, wheat planted in a prepared seedbed has little chance to become fully established before the onset of cold weather. Another technique for wheat establishment is to overseed, using an airplane or highboy while the soybean foliage still makes a full canopy and use the falling leaves as a mulch (Sanford, 1979).

In Mississippi work, Sanford (1979) showed that wheat yield and hence germination was better with a full canopy than with a partial one. Results are shown in Table 10-5. His results indicate that the canopy provides shade which in turn assures a moist surface soil for germination of broadcast seed.

In sandy surface soils which dry out quickly, overseeding wheat in North Carolina either in the Piedmont or Coastal Plain was quite unsuccessful (Lewis and Phillips, 1976), yielding only 25–40 percent as much wheat as seeded conventionally after soybeans and lowering the soybean yield as well (Table 10-6). Clearly, soil characteristics are important in determining the most effective way to plant a small grain in a multicropping system.

Weed Control

Control of weeds is vital and, because of timing and herbicide incompatibility problems, it can be a greater challenge in multiple cropping.

Table 10-5. Yield of wheat overseeded in soybeans depending on percent canopy. (Reproduced with the permission of Sanford, from *Agronomy Journal* 71:109–112 [1979].)

OVERSEEDED IN BEANS	WHEAT YIELD TWO-YEAR AVERAGE	SOYBEAN YIELD TWO-YEAR AVERAGE
	(mt/ha)	
With 76% canopy	2.7	2.2
With 100% canopy	3.2	2.4

Table 10-6. Wheat and soybean yields with conventional (C) and overseeded (OS) wheat followed by no-till soybeans. (Reproduced from Lewis and Phillips, 1976.)

	REGION			
	PIEDMONT		COASTAL PLAIN	
	WHEAT YIELD	SOYBEAN YIELD	WHEAT YIELD	SOYBEAN YIELD
	(mt/ha)			
Wheat (C) Soybeans (NT)	2.4	2.0	1.9	2.0
Wheat (OS) Soybeans (NT)	1.0	1.7	0.4	1.5

Curiously enough, however, the sequential change in crops gives the grower chances to control some problem weeds more effectively than if only one crop was grown. An example of this is the corn-wheat-soybean sequence (three crops in two years) versus continuous corn.

With corn, Johnsongrass is difficult to control, even with conventional tillage and the use of herbicides. When soybeans are grown, there are a number of effective control measures against Johnsongrass, starting with an incorporated herbicide such as Treflan before planting and including rope-wick wipers containing Roundup when soybean plants are large. Perennial vines can be attacked with Roundup as well.

The most common fear of herbicide injury from one crop to another is that of Triazine used on corn and their suppression of small grain growth. Triazine injury to wheat is minimal when it is applied at recommended rates, at least in the southeast and in the southern corn belt. In addition, under no-tillage, the Triazine activity decreases much faster than under conventional tillage (Kells *et al.*, 1980). This effect appears to be due to two causes: (1) the organic layer at the surface and its binding of the triazines, and (2) the lower surface pH with no-tillage than with conventional tillage.

It is certain that with multiple cropping of any sort, competition from plant growth and herbicides will be more important in controlling weeds than they are in monocropping. This is the case since opportunities for tillage are much reduced. In the system wheat or barley followed by soybeans, corn or sorghum, it is absolutely vital that a thorough job of weed control be done in the brief time available between the first and second crop. Often, in clean wheat fields, the stubble looks as if it would suppress all weed growth without further treatment. A close look will reveal that this is not true, for small etiolated weeds will spring out rapidly after wheat harvest. In addition, volunteer wheat can be a problem. This is a critical time to use both a contact and a residual herbicide, especially if no mechanical tillage is to be used.

Fertilizer Practices

The requirement for plant nutrients varies greatly according to crops grown and plant parts removed from the soil. For example, with wheat for grain followed by soybeans for grain, the fertilizer requirements are low relative to most cropping systems. Wheat requires 70–100 kg/ha of N and a soil that is relatively high in phosphorus and medium in potassium. Soybeans require no fertilizer nitrogen and only medium levels of P and K. In contrast, one of the most demanding cropping systems in terms of fertilizer is wheat silage followed by corn silage. The first crop requires 70–100 kg/ha of N and high K, and the corn silage requires 150–225 kg/ha of N and high P and K (Murdock and Wells, 1978). Total K removal in the double crop silage can be 350 kg/ha. For comparison, a crop of corn for grain removes less than 115 kg/ha. It is apparent that the nutrient removal by the double crop of wheat-soybeans is close to that of corn, whereas the double-cropping system of wheat silage-corn silage has much higher nutrient removal, especially of K. Hence, it is impossible to generalize about soil fertility requirements for multiple cropping as opposed to a single crop.

Equipment

The equipment required for multiple cropping on a given farm usually is exactly the same as that required for traditional cropping systems. For example, if a farmer has been following a corn, soybean rotation, he can, without any additional equipment purchase, follow a corn, wheat-soybean or a corn, wheat-sorghum multiple cropping scheme. Similarly, a dairy farmer with silage-making equipment can change from corn silage to wheat silage-corn silage with no additional purchases.

The ability to produce additional crops without the purchase of new equipment is, in fact, one of the attractive features about multiple cropping as practiced in the U.S. It makes production schemes quite flexible by making adjustments to equipment already on hand. For example, row width adjustment on a planter is the major change between soybeans and corn. It appears doubtful that any multiple cropping scheme that requires additional equipment will find favor with farmers, at least under present economic conditions.

Management of Multiple Cropping

Multiple cropping is an attempt to use the same resource base of land and equipment to take advantage of additional sunlight and water. As this demands more operations during a calendar year and more decisions with regard to herbicides, insecticides and fertilizers, it more effectively utilizes the farmer's time. A farmer who is permanently behind time operating a single cropping system would be well advised not to try multiple cropping because of the additional chores that will evolve upon him. A farmer with management ability, on the other hand, has prospects of improving his financial situation, both by increasing his gross income and by providing additional "insurance" against crop and/or price deficiencies.

Areas of management requiring attention in a multiple cropping system including the following:

1. Machinery: its adequacy to do the job and its repair.
2. Herbicides and other chemicals: their effectiveness and their effect on succeeding crops.
3. Keeping up with crop progress in individual fields so that harvesting and planting is as timely as possible.
4. Storage and drying for crops.
5. Marketing for best price and to assure cash flow.
6. Awareness of changing conditions in all these subjects.

Of these management areas, the one which stands out in double cropping is Number 3 – keeping harvesting and planting as timely as possible. Given substantial yield losses for each day of delay, the biggest factor in production (controlled by the farmer) is most often timeliness. For example, if the losses of soybean yields after a given date are 62–75 kg/ha-day, the cash losses on a 200-ha planting of soybeans would be as much as $4,000–$5,000/day. A delay of one week probably would make the difference between profit and loss.

Harvesting wheat too early, on the other hand, might increase drying costs so much that profitability of that crop is lost. Excessive rain or lack of rain also will influence the farmer's decision on when to harvest and plant. Managers who can integrate all these factors turn out to be successful, while those who do not go into other types of work.

Profitability of Multiple Cropping

Despite other attributes of multiple cropping, if it does not, over a period of time, provide more net income to the farmer, it will not be practiced. The rather limited figures available show that a double crop of wheat-soybeans is only slightly more profitable than a single crop of corn (Lewis and Phillips, 1976). If it were assumed that corn and wheat-soybeans were exactly the same in yearly profit potential, it is still likely that on the long term it would be more profitable to have some of the farm in the wheat-soybean double crop each year. The major reasons for this are hedging, cash flow and weed control. In the case of hedging, the prices of soybeans and wheat are gambled against the price of corn, a common practice even without double cropping.

Possibly the most intriguing feature about multiple cropping as related to profitability is the possibility that it offers to improve cash flow on the farm. An example may serve to illustrate the idea. If a farmer has borrowed $75,000 to plant 200 ha of corn and 200 ha of soybeans, and he is paying an interest rate of 18 percent from May through October, his interest cost is $6,750 for the six-month period. If he were growing wheat-soybeans rather than the soybeans alone, and the yields of wheat were 2.9 mt/ha, he could pay off the principal in July, save $3,375 in interest and still have some change left. The increase in profitability due to saving loan interest often was not considered when interest rates were lower, but it can be an important factor under present (and probably future) conditions. Table 10-7 shows data for corn versus wheat, soybean and wheat-soybean gross returns using August 25, 1981 cash prices, Chicago. It is obvious that the double cropping system becomes more attractive

Table 10-7. Gross returns on corn and double cropped wheat and soybeans at selected yield levels.

Corn	mt/ha	6.8	7.5	8.2	9.5
	$/ha	$722	$795	$867	$1012
Wheat	mt/ha	2.0	2.7	3.4	4.1
	$/ha	$272	$363	$454	$545
Soybeans	mt/ha	2.0	2.4	2.7	3.4
	$/ha	$484	$565	$646	$808
Total	$ value from double crop wheat-soybeans	756	928	1000	1353
	$ value from corn	722	795	867	1012

relative to corn as yields of all three crops increase. This illustrates that multiple cropping has to have good yields to compete favorably with a monocrop such as corn. Actual results from Oklahoma (Crabtree and Rupp, 1980) show similar wheat and soybean values.

Second crops other than soybeans have difficulty in competing very well with corn. Possibilities for multiple cropping in colder and drier areas include sorghum and sunflower. Their use will be partially dependent on their profitability relative to corn. Data on profitability of sorghum as a second crop in the humid region show discouraging results (Lewis and Phillips, 1976).

Perhaps one of the most urgent reasons that wheat-soybeans are favored in a rotation with corn has to do with weed control. This effect is not likely to be noticed in any given year but has a long-term benefit. In fields where Johnsongrass and/or shattercane are prevalent, control with continuous corn is difficult. With soybeans, in contrast, there are both preplant and post plant control available. Preplant incorporation of Treflan does not fit in well with a double cropping system, at least in some years. Late treatment with a rope-wick using Roundup herbicide is widely used in soybeans, a practice not possible in corn due to its tall growth habit. On the other hand, broadleaf weeds, which compete well in soybeans, do not compete as strongly with corn, so that the rotation is favorable for both crops.

In summary, profitability largely determines the type of multiple cropping that will be practiced. At the present time, soybeans appear most attractive for areas where they can be grown. Outside these areas, either due to drought or a short growing season, sorghum and sunflower appear to have promise. Winter crops other than wheat or barley would include rye for grazing and cover and winter legumes such as hairy vetch for nitrogen and cover.

THE FUTURE OF MULTIPLE CROPPING

The vast array of multiple cropping systems that possibly can be grown makes it very difficult to predict the number of hectares and areas of the U.S. that will be affected by this practice. Therefore, in the sections that follow, most emphasis will be placed on proven systems.

To generalize types of multiple cropping, it will be divided into winter-summer double cropping, summer-summer double cropping and winter-summer-summer triple cropping. Included in some of these systems is so-called "relay planting," where the second crop is introduced before the first crop is harvested.

Winter-Summer Double Cropping

As has been indicated, it is anticipated that the major system of winter-summer double cropping will continue to be wheat-soybeans. There are several reasons for this prediction. In the first place, it has a large number of hectares at present and is growing rapidly. Second, it probably will continue to be the most profitable combination in areas where it can be grown. Third, farmers know how to manage this cropping sequence. And fourth, it is highly compatible with corn, especially under maximum or no-tillage. Fifth, world acceptance and market availability for wheat and soybeans has long been established.

The potential of number of hectares for wheat-soybeans is high in the warm-humid region of the U.S. (south of 40° latitude and east of 75-cm annual rainfall). At the current time, only 4,400,000 ha of wheat are grown compared with 12,000,000 ha of soybeans. It is not known what proportion of wheat is followed by soybeans in the entire region. In Kentucky, essentially all the wheat is followed by soybeans and nearly all is planted using no-tillage or minimum tillage. The number of hectares of soybeans is nearly static but the number of hectares of wheat has risen rapidly, indicating that, on well-drained soils at least, the multiple cropping system of wheat-soybeans makes up as much as one-third of the soybean hectares. This is expected to grow, perhaps to double, within the next ten years or so. Its limit is governed by wet soils on which wheat cannot be grown successfully and by economic factors.

The systems wheat-sorghum and wheat-sunflowers may well become accepted where soybeans are limited by dry and/or cold weather. It is not expected that these systems will compete well in locations where soybeans will yield in the 2.0 mt/ha range. Use of valuable irrigation water for a second crop of sorghum or sunflower is not very likely except in emergencies. Use of irrigation water for soybeans will take precedence, so that even in some dry areas (southern and southwest United States, for example), soybeans will be favored as a crop.

Brazilian livestock-grain farms use annual rye grass for sheep and cattle grazing, allow the rye grass to mature and drop seed, dessicate the rye grass and plant soybeans by no-tillage. This system provides high quality forage without reseeding the rye grass during the cooler period of the year and a cash grain crop with soybeans for grain sorghum. Land with 12–15 percent slope is used without erosion problems, enhancing land use and income.

Summer-Summer Double Cropping

As has been pointed out previously, in areas where there is a growing season of perhaps 250 days or more, two corn crops or a corn-sorghum sequence can be considered if the first crop is established as early as possible. Unfortunately, in the continental U.S., at least, long growing seasons are associated with elevated temperatures, which severely depress the second crop yields in corn (Widstrom and Young, 1980). Perhaps a more logical sequence to try, given water limitation and high temperatures, is corn-soybeans, soybeans-soybeans or sorghum-ratoon sorghum. In all these systems, however, the risk of crop failure to the second crop is fairly great unless supplementary irrigation is available. Considerable investigatory work has been done on these systems, but none of them appear to have been accepted by farmers.

Of these systems, the one which appears to be most commercially viable is corn-soybeans. It has several advantages, including maximum gross income and lower costs for the second crop. Applying expensive nitrogen to a doubtful second crop such as corn increases the risk of financial loss. Additionally, because soybeans bloom over a longer period of time, their yields tend to be hurt less by short periods of drought during flowering. Corn, on the other hand, requires excellent conditions during silking and tasseling, or else yields will be low. For all these reasons, it appears likely that soybeans would make a better second crop than would corn. This system has a potential on at least a million hectares of land in the deep south and southwest.

Sorghum-sorghum (second crop as a ratoon) also has promise as a system except that total returns tend to be low compared to corn-soybeans. It may well become a system adapted to areas where double cropping can be done in wetter than average years but where yearly double cropping is impossible. The system could be managed so that no nitrogen fertilizer is applied unless sufficient water is available to pay for the fertilizer. Then, fertilization can be carried out late with some confidence.

Winter wheat followed by grain sorghum followed by corn in the same year is practiced in Argentina and other tropical regions.

Winter-Summer-Summer Double Cropping

In Florida and other Gulf Coast states, there is a potential of adding a winter vegetable crop to two summer crops, or following a winter-summer sequence with a late fall planting of a cool-season vegetable. This type of system has the advantage of producing the vegetable crop when prices are relatively high, and still producing field crops competitively with the rest of the nation. Systems such as these are common already in the lower Rio Grande Valley of Texas and they seem to have developed mostly to take advantage of both the possible high prices of vegetables and the stability of field crops. However, the total number of hectares in this type of system will be rather low and insignificant in terms of total grain production. A system with small grain followed by two summer grains will be even less prevalent.

Tropical regions have the potential to grow winter wheat, soybeans and grain sorghum rotations each year.

Other Multiple Cropping Systems

As the price of nitrogen fertilizer continues to increase, reflecting rising natural gas prices, the tendency to produce a significant portion of this nitrogen with legumes probably will increase considerably unless grain crop prices rise accordingly. Such a rise in grain prices is neither likely to occur nor last for a long period of time, given past history.

In addition to conventional rotations with forage legumes, there appears to be promise for winter legumes in corn production supplying most but not all the required nitrogen. Studies in Delaware, Virginia, Georgia and Kentucky show hairy vetch, overseeded or drilling after corn, to be the most consistent performer in terms of both cover and nitrogen fixation. Results from Kentucky are given in Table 10-8. Schemes of this type probably will find increasing use in response to increasing nitrogen prices in areas where the legumes can survive the winter.

Another instance of using legumes, this time a true intercropping, is "renovation" planting of red clover in grass, both to improve quality of pasture and to provide residual nitrogen for the grass. This system is rather widely used from Pennsylvania through Missouri, and would be used more intensively if livestock prices were not so consistently unfavorable compared to crop prices. Other examples of intercropped pastures are alfalfa in grass and winter grasses or small grains in dormant bermuda grass.

External Factors Affecting Profitability of Multiple Cropping

At the present time, the future of several types of multiple cropping looks favorable. Multiple cropping is favored by expensive land, expensive energy supplies, high interest rates, cheap cattle prices and relatively stable grain markets. However, drastic changes in any of these factors could change the outlook for multiple cropping very quickly. For example, a world grain glut would have a negative effect on multiple cropping of grains because the marginal profits possible with multiple cropping would disappear. Expensive cattle would compete for all crops in the areas where multiple cropping is most generally practiced, i.e., the fringes of the corn belt. In these

Table 10-8. Yield of corn following corn and following hairy vetch in Kentucky. (From Frye, 1981.)

SURFACE RESIDUE	N RATE	CORN YIELD
	(kg/ha)	(mt/ha)
Corn	0	5.2
	50	5.6
	100	7.1
Corn-vetch-corn	0	6.6
	50	7.6
	102	8.9

areas (Kentucky and Missouri are good examples), cattle are almost always present on farms because some of the land is unsuitable for cropping. If cattle carry the possibility of higher profits, more and more of the farm becomes "unsuitable" for cropping.

A fall in interest rates also would harm multiple cropping because the need for cash flow in slack time would not be so great. Similarly, a fall in land prices would oppose multiple cropping. No predictions are made about any of these factors. It is clear that multiple cropping is a response to economic conditions rather than an idealistic effect to produce more food.

Combining Agronomic Practices for Multi-Cropping on Sloping Land

The rapid increase in soybean production in the lower midwest and southeastern U.S. has created or increased a serious erosion problem. Wells (1979) has combined, tested and tried agronomic practices into a workable no-tillage system that will permit soybean production on land with slopes in the 3–9 percent category without noticeable soil erosion. The combination described makes use of (1) the standard midwestern corn-soybean rotation; (2) seeding a small grain for winter cover; (3) double cropping; (4) no-till planting; and (5) contour strip cropping. In addition to these basic components, sod waterways would serve as pathways for slope runoff of excess rainfall. This system would make use of residue and would provide good erosion control, since the field would never be cultivated except for a fall discing, and both corn and soybeans would be no-till planted. The weak link of this aspect is obtaining enough fall growth of wheat to provide good over-winter cover. With "normal" fall weather in this latitude we usually can complete soybean harvest by mid to late October, and seed wheat immediately thereafter. Advent of abnormally cold weather by early November can depress early wheat growth enough to provide inadequate cover. Two safety valves could also be added by establishing and maintaining a sod strip (just wide enough to mow) between each crop strip, or in the event that harvest is too late to justify seeding the small grain, simply leave the strips untilled over winter. In this case there will be adequate corn residue to protect that strip, and there will be partial protection from what soybean residues remain on soybean strips. The worst which would likely happen under these circumstances would be sheet erosion across soybean strips which would stop as it reached the next corn strip down hill but would allow continued long-term production of soybeans on sloping land and would allow the use of sloping land to produce grain intensively while controlling erosion. In areas where small grains can be fall planted, this system would result in each ha used producing 1.5 ha of harvested crops (0.5 ha soybeans, 0.5 ha corn and 0.5 ha small grain). It would also result in better broad-spectrum weed control than that possible in fields used continuously for either crop alone.

Excellent yields of corn and soybeans have been recorded in Table 10-9. This package is being rapidly adopted by innovative farmers in Kentucky. It has proven to be effective in erosion control and much less expensive than many engineering structures.

Table 10-9. Yield of corn and soybeans grown in strip rotation. (From Wells, 1979.)

STRIP	CROP	1977	1978	1979	THREE-YEAR AVERAGE
				(mt/ha)	
1	Corn	9.7	9.2	9.4	9.4
	Soybeans	4.3	3.3	2.9	3.5
2	Corn	10.0	8.8	10.0	9.6
	Soybeans	4.6	3.5	3.3	3.8
3	Corn	9.7	8.5	9.7	9.3
	Soybeans	3.4	3.0	3.7	3.4
4	Corn	9.4	8.6	10.1	9.4
	Soybeans	3.6	2.9	3.7	3.4
Average	Corn	9.7	8.8	9.8	9.4
	Soybeans	4.0	3.2	3.4	3.5

REFERENCES

Allen, L. H. Jr., T. R. Sinclair, and E. R. Lemon. 1976. Radiation and microclimate relationships in multiple cropping systems. In *Multiple Cropping*. R. I. Papendick, P. A. Sanchez, and G. B. Triplett, Editors. (Am. Soc. Agron. Special Publ. No. 27), pp. 171–200.

Crabtree, R. J. and R. N. Rupp. 1980. Double and monocropped wheat and

soybeans under different tillage and row spacings. *Agron. J.* **72**:445–448.
Frye, W. W. 1981. Legume cover crop provides nitrogen for no-tillage corn. *No-Till Notes 1, No. 2.* University of Kentucky.
Kells, J. J., C. E. Rieck, R. L. Blevins, and W. M. Muir. 1980. Atrazine dissipation as affected by surface pH and tillage. *Weed Sci.* **28**:101–104.
Lewis, W. M. and J. A. Phillips. 1976. Double cropping in the eastern United States. In *Multiple Cropping.* R. I. Papendick, P. A. Sanchez, and G. B. Triplett, Editors. (Am. Soc. Agron. Special Publ. No. 27), pp. 41–50.
Murdock, L. W. and K. L. Wells. 1978. Yields, nutrient removal and nutrient concentrations of double-cropped corn and small grain silage. *Agron. J.* **70**:573–576.
Sanford, J. O. 1979. Establishing wheat after soybeans in double-cropping. *Agron. J.* **71**:109–112.
Slack, C. H., W. W. Witt, and R. M. Bullock. 1980. Herbicide evaluation trials – 1980. University of Kentucky, Department of Agronomy.
TeKrony, D. M., A. Phillips, and T. Howard. 1976. The effects of swathing wheat on the date of harvest, yield and seed quality. *Agron. Notes Vol. 9 (No. 2), February.* Cooperative Extension Service, University of Kentucky.
Unger, P. W. and B. A. Stewart. 1976. Land preparation and seedling establishment practices in multiple cropping systems. In *Multiple Cropping.* R. I. Papendick, P. A. Sanchez, and G. B. Triplett, Editors. (Am. Soc. Agron. Special Publ. No. 27), pp. 255–273.
Wells, K. L. 1979. Southeastern Soil Erosion Control and Water Quality Workshop. Agronomy Notes, University of Kentucky.
Widstrom, N. W. and J. R. Young. 1980. Double cropping corn on the coastal plain of the southeastern United States. *Agron. J.* **72**:302–305.

11
Equipment

Shirley H. Phillips
Associate Director of Extension
College of Agriculture, University of Kentucky

PLANTERS AND PLANTING NO-TILLAGE

The equipment requirements for no-tillage are as varied as those for conventional systems with influencing factors that include power source, human, animal or internal combustion engines; size and scale of operations; species of crops grown; cropping sequence; topography of land resources; soil types; vegetation and residues present; timing of planting and harvest; multicropping systems used; availability of equipment; and climatological variation and conditions.

Minimum equipment required for no-tillage agriculture will consist of a planter, herbicide applicator and harvester. Several types of planters are available for powered mechanical planting. The no-tillage planting equipment must be capable of penetrating mulches and to operate in varying soil conditions. Generalization can be drawn on planters with requirements as follows: sufficient weight to penetrate mulch and soils; open furrow or punch into soil with sufficient width and depth for optimum seed placement; meter different size seed at varying spacing; adequate coverage of seed with soil; multi-tool bar capacity for various coulter configurations; and rolling coulters to prevent residue buildup in front of planters.

Present planters fall into six major categories: (1) chisel type, (2) angled coulters, (3) passive rolling coulters, (4) powered blades, (5) rotary strip tiller and (6) combinations of these. (Figure 11-1.)

Several passive coulters are in use today which have shown good performance. (Figure 11-2.)

When residues are wet, especially small grain and rice straw, clean cutting may require special coulters and configurations to slice through mulches followed by the coulter to produce the narrow tilled band of soil. (Figure 11-3.)

Planter selection may be enhanced by considering the advantages and disadvantages of each major type.

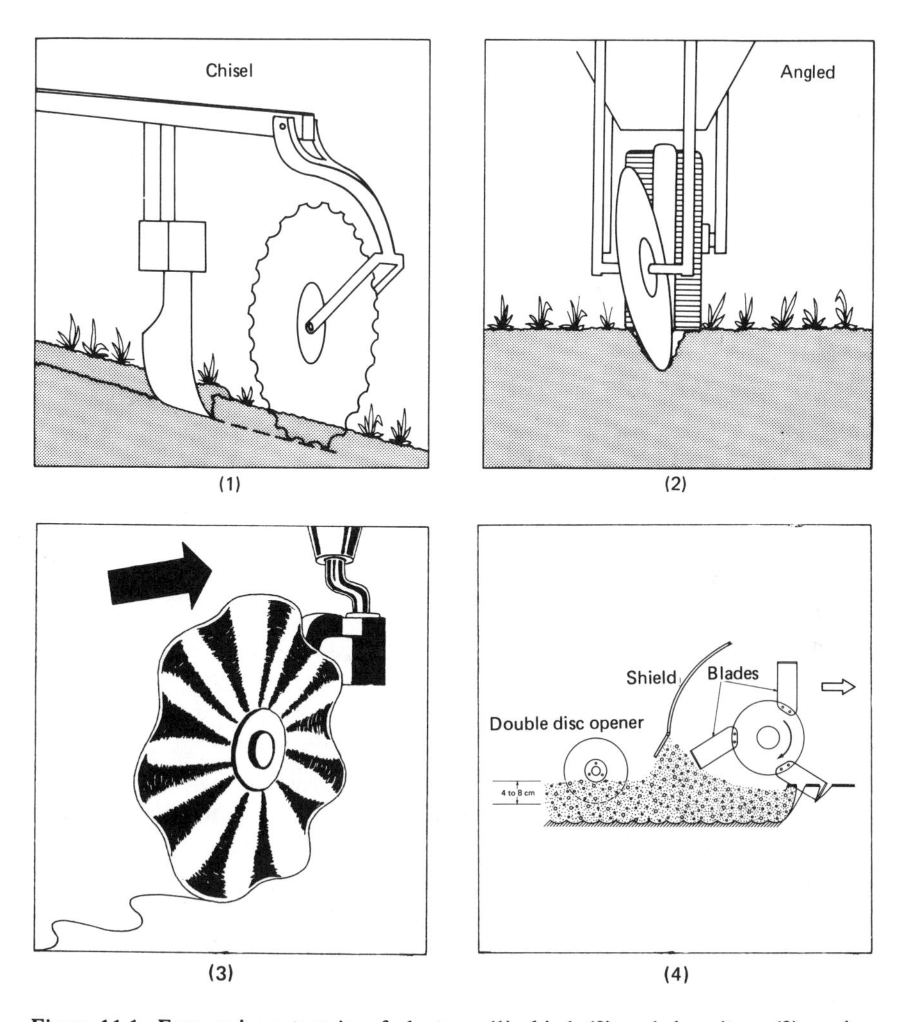

Figure 11-1. Four major categories of planters: (1) chisel; (2) angled coulters; (3) passive rolling coulters; (4) rotary strip tiller.

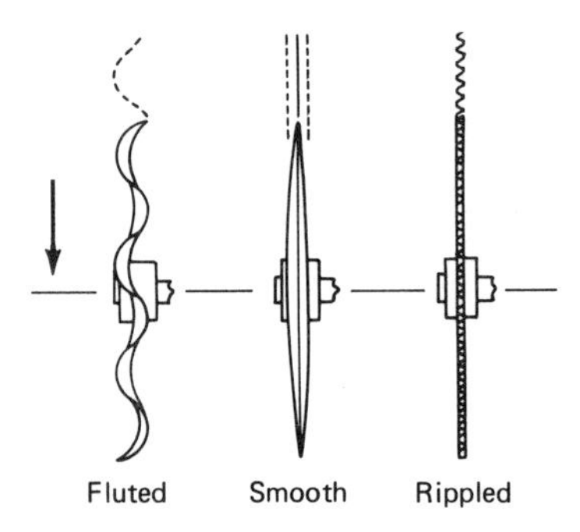

Figure 11-2. Types of passive coulters; (1) fluted; (2) smooth; and (3) rippled.

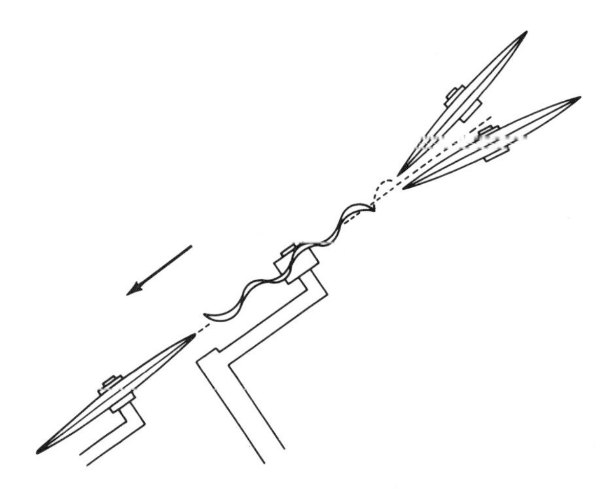

Figure 11-3. Sketch showing a 3-bar planter with smooth coulter in front followed by fluted coulter and planting unit.

Types of Planters

A variety of effective "minimum tillage" and "till-plant" planters are available. They include planters with chisels, sweeps or rotary strip tillers. Power requirements for most of these implements are higher than for no-tillage planters. Detailed descriptions of minimum-tillage equipment are not included in these discussions because by definition and performance this equipment manipulates the soil too much to be classed as "no-tillage" machinery.

For several years the greatest factor limiting adoption of no-tillage was lack of a commercially available planter that would place seed properly in varying soil, moisture, and crop residue conditions. The earliest no-tillage planters were usually existing conventional models modified by pioneering farmers.

A no-tillage planter must fulfill several basic requirements. It must:

1. Be heavy enough and strong enough to plant under adverse soil conditions and cut through previous crop residues.
2. Provide a narrow band of tillage for receiving the seed. A 5–8-cm-wide and 7–15-cm-deep soil manipulation will be fully adequate, according to the experiences of successful no-tillage farmers and investigations of agricultural engineers and crop specialists.
3. Plant seed at different depths. Seed size, soil temperature and depth to adequate moisture dictate depth of seed placement. For example, soybeans planted late may need to be placed deep enough to be in contact with soil moisture to assure good germination. Accurate control of planting depth from 2.5–7 cm will be required.
4. Cover and firm soil around seed. Usually soil for seed coverage is prepared by the row opener and seed coverage accomplished by the packer wheel. Coverage is important to assure germination and provide protection from bird and rodent damage. Firming of the soil in the row is needed to reduce air pockets and to maintain desirable moisture conditions around the seed.

At the present time there are a number of commercially available no-tillage planters of the types listed above. Each of the types have special advantages and certain drawbacks. A grower evaluation would list the strong and weak points as follows.

Chisel Type No-Tillage Planter

Advantages:

1. Penetrates both heavy soils and soils in poor physical condi-
1. tion.
2. Depth of chisel opener and seed drop easy to adjust.
3. Relatively lower cost per row unit.
4. Available in three-point hitch.
5. Can apply starter fertilizer below surface.

Disadvantages:

1. Does not work well in stones, stumps, roots or other soil obstructions.
2. Trash may accumulate in front of the chisel, reducing ha/day planted.
3. Soil moisture conditions need to be favorable. When the soil is very dry and firm, this type planter tends to pull up chunks of soil, or to make too many cracks in the soil near the row, allowing for more rapid drying.
4. Produces air pockets and tends to leave a small tunnel in seed zone when planting under wet soil conditions.
5. Requires more power than the other types of planters.
6. Leaves fields rougher than other types.
7. Requires close attention to planter unit adjustment on uneven fields.

Angled Coulter No-Tillage Planter

Advantagcs:

1. Penetrates heavy sods, thatch or crop residues.
2. Lower power requirements than chisel type planter and similar to other passive coulters.
3. Relatively low weight and cost.
4. Available in three-point hitch.
5. Attains better seed coverage than chisel type planters.

Disadvantages:

1. At present stage of development, row fertilizer may be applied at surface only.
2. Requires careful adjustment of coulters and planting units.

Passive Rolling Coulter

Advantages:

1. Greater versatility in adding weight and in making modifications to fit various planting conditions and row widths than other two types.
2. Allows use of small grain, grass and legume seeding units.
3. Penetrates greater depths of residues than other types.

(a)

(b)

Figure 11-4. Photograph of (a) fluted coulter no-tillage planter (courtesy of Allis Chalmbers Co.) and of (b) ripple coulter planter (courtesy of John Deere and Co.).

4. Plants satisfactorily under wetter soil conditions than the other two types described.
5. Slightly higher planting speed possible than with chisel planters.
6. Fertilization placement below surface is possible.
7. Drawbar power requirements per row unit generally the lowest of currently available commercial models.

Disadvantages:

1. Costs slightly higher.
2. When planting shallow, seed coverage in some soils may be difficult if planter is not properly adjusted.
3. Requires careful attention to properly aligned coulters and planter unit adjustments.

Powered Blade

Advantages:

1. More uniform planting.
2. Slices through residues effectively.
3. Operates in wide range of soil types.
4. Planters can be lighter weight.
5. Improved emergence and stands.

Disadvantages:

1. Requires more power.
2. Higher fuel and energy requirements.
3. Increased cost of maintenance and repairs.
4. More moving parts.
5. Problem of planting on wetter soils.
6. More dust in seeding operations.

Band Roto Tiller

Advantages:

1. Destroys weed present.
2. Prepares easily managed tilled area for seed placement.
3. Increases soil temperature in seed zone (temperate regions).

Disadvantages:

1. Extremely high power and energy requirements.
2. Soils must be relatively dry and workable.
3. Increased weed problems in later stages of crop development.
4. High maintenance.
5. Deteriorates soil structure.
6. Operational problems in heavy mulches.
7. Higher soil erosion possibility.
8. Increased evaporation of soil water.

A new generation of no-tillage drills have appeared for soybean, small grain, rape, flax and other small seeds with row spacing capabilities not afforded by earlier models used in corn and soybean plantings. (Figure 11-5.)

No-tillage planters are usually higher in price than conventional pesticide applicators become more standard than needed with conventional units. The importance of learning new technology mentioned in the introduction can be demonstrated in the planter operation. Soils that are friable and with ideal moisture at planting require less manipulation of equipment and technical knowledge on the part of the grower. The configuration of the coulters, weight of the planters and other factors become more important in soils with less desirable planting conditions. As soils become dry or on soils with poor physical condition, additional weight over the coulter may be required for adequate penetration and tilling action of the coulters. When these conditions prevail, 10–15-cm penetration with coulter(s) aligned with the planting unit is most desirable for good seed incorporation and coverage. (Figure 11-6.)

It is possible to plant no-tillage when soils are wet to the point where plowing or discing would be impractical. Wet soil conditions demand regulating coulter penetration at the same level or slightly higher than the planter unit.

Should proper seed coverage continue to be a problem because soil does not shatter and fall into place slightly off center, misalignment of the coulter and planter unit may assist in correcting this problem. (Figure 11-7.)

(A)

(B)

Figure 11-5. Photograph of (a) haybuster seeder (courtesy of Haybuster Manufacturing Co.) and (b) power-till seeder (courtesy of John Deere and Co.).

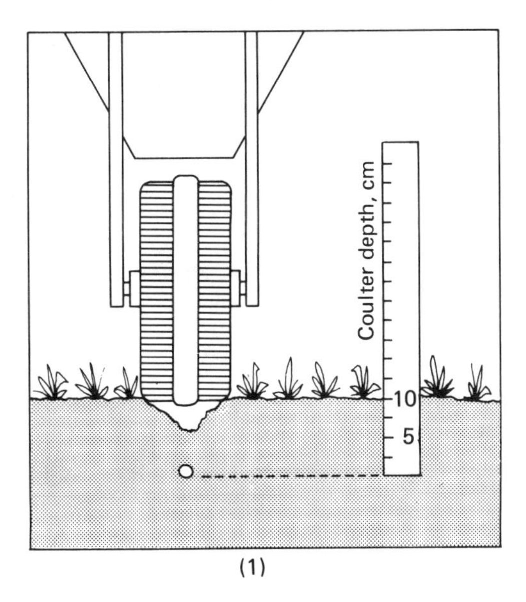

(1)

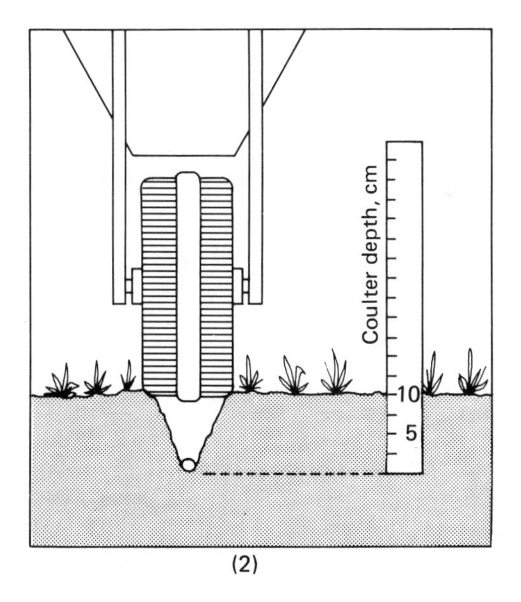

(2)

Figure 11-6. Sketch showing alignment of coulter, seed placement and press wheel: (1) proper coulter depth and soil coverage of seed; (2) coulter penetration slightly less than planting unit and seed at bottom of coulter silt without proper soil coverage of seed.

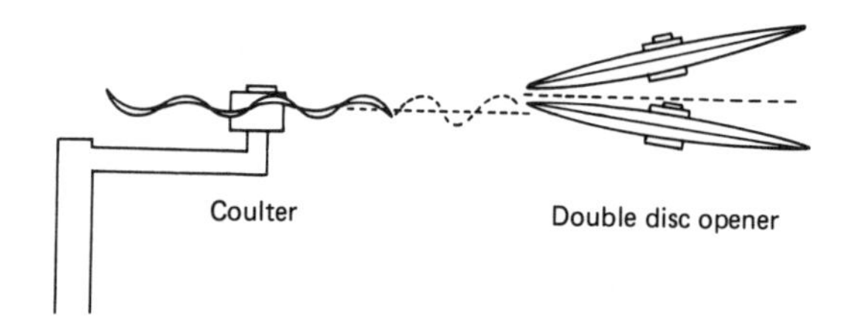

Figure 11-7. Sketch showing coulter and planting unit opener off-center

Tractor Requirements for No-Tillage

The horsepower requirements per ha are reduced by 30–50 percent on no-tillage farms. This reduction may be in the form of smaller tractors required. Most tractors in the 60–80-horsepower range can accommodate up to 8-row equipment. The most noticeable limitation is not in drawbar power but sufficient hydraulic system to lift the planter unit for transport and turns while planting. Longevity of the tractor is improved by eliminating plowing and discing, two operations requiring the most power over extended periods of time. The combination of reduced tractor numbers and decreased depreciation can affect net income significantly. Farmers with equipment sized for conventional tillage of crops will realize less immediate savings when converting to no-tillage because of existing machinery inventory. The saving in this situation will be recognized over an extended period of time. Elimination of plows, discs and cultivators is possible with no-tillage and will further decrease needed implement inventory. Farmers using combinations of both no-tillage and conventional methods will realize less difference in size and number of implements.

Pesticide Applicators

Today's modern crop production systems usually require equipment to meter and apply pesticides. It will be impossible to produce crops using no-tillage techniques without these application devices. (Figure 11-8.) Herbicide application discussed in Chapter 7 will provide application information in detail. It is imperative that equipment capable of meeting those requirements and the technology associated with this equipment be in place before moving into this system. This equipment may be mounted on planters for one-trip planting or equipment

Figure 11-8. Photograph of no-till planter with insecticide applicator (courtesy of John Deere and Co.).

may be selected for post planting and preemergence application of pesticides in separate operations.

Farmers with equipment that have weed and pesticide application operations monitored by electronic devices tend to combine all operations into a one-trip over the field system to further reduce crop establishment time and conserve energy. Others plant, apply insecticides and follow with herbicide application in a separate operation.

Hand Planting

Wijewardene (1979a), and others with the International Institute of Tropical Agriculture (IITA), Ibadan, Nigeria, has made many comparisons that show great promise for no-tillage to assist small farmers in the humid tropics. This group recognized the impossibility of passive multi-counters being pulled by humans. IITA's mechanization of the "hand job" planter into a rolling punch planter unit has made no-tillage operations possible in developing countries (see Figure 11-9). This planter reduces the time to less than half that of the single job planter which was significantly faster and more efficient than hand tools such as the machete or wooden spike. The rotary punch planter developed there is composed of a series of five to six pointed jaws on a periphery of a disc rolled over the soil area to be planted. The jaws or protrusions penetrate through the mulch into the soil, then opened at the lowest position to place seeds before the jaws emerge from the soil. A compaction wheel to press soil over the seed follows the rotary disc planting unit.

Planting units varying from one to four rows can be used depending on the needs and size of the farmer's operation. Thus, energy and time can be reduced.

Figure 11-9. Photograph of a hand, rolling-punch planter (IITA planter).

Table 11-1. Energy and time required to establish corn in conventional and no-tillage systems under large scale tropical farming conditions. (From Wijewardene, 1979b.)

TREATMENT	NUMBER OF PASSES	ENERGY (megacal/ha)	LABOR (hr/ha)
1. *Conventional*–disc plough, disc harrow twice, and plant	4	56.1	5.4
2. *No-tillage*–mow, spray herbicide, and plant	3	12.4	2.3

The limiting factor in producing food on small farms typical of the humid tropics is in hand digging and preparation of the soil prior to planting. An excellent comparison (Table 11-2) was shown in research at the International Institute of Tropical Agriculture, Ibadan, Nigeria (I.I.T.A.) in favor of no-tillage. No-tillage tools were all hand held; i.e., controlled droplet applicator (CDA) sprayers for herbicide and insecticide, and I.I.T.A. auto-feed "punch" planter. Yields were

Table 11-2. Manpower requirements for no-tillage and conventional crop establishment systems for maize and cow pea on Savannah (imperata cylindrica) covered land under small-scale (manual) farming conditions. (From Wijewardene, 1979b.)

OPERATION	NO-TILLAGE	CONVENTIONAL
	(man-hr/ha)	
Field Preparation		
a. Slash, burn and till manually		180
b. Controlled droplet applicator (CDA) spraying with systemic herbicide	5	
Seeding		
a. Manual planting into tilled soil with machete (low plant population)		20
b. Auto-feed "punch" planting (maize-cowpea 75 x 25)	30	
Weed Control		
a. Manual weeding twice		280
b. CDA spraying with preemergent herbicide	5	
Fertilizer Application		
a. Banding by hand along rows		25
b. Using a hand propelled band applicator	6	
Plant Protection		
a. Knapsack spraying of insecticide		10
b. CDA spraying of insecticide	2	
Total man-hours spent to establish crop on one hectare of land (not including harvesting)	48	515

significantly higher on the no-tilled area due to better weed control and plant population.

Rice production provides a large supply of food and is favored over other food grown in much of the world. The rotary punch planter can be used in this important food crop with similar reduction in

Table 11-3. Comparison of machine and man hours by systems and tools used for alternative systems of upland rice establishment. (From Wijewardene, 1979b.)

SYSTEM AND TOOLS USED		MACHINE	MAN
		(hr/ha)	
System 1. Small Tractor and Improved Tools			
Ploughing (hand tractor)		16	32
Harrowing and leveling (hand tractor)		8	16
Seeding (two-row "pull" seeder)			42
Weeding (inter-row, hand-weeder)			76
Insecticide spraying (CDA, twice)			1
	Totals	24	167
System 2. Small Tractor Plus Herbicide			
Rotary tilling (hand tractor)		13	28
Seeding (broadcast)			3
Light harrowing (hand tractor)		4	8
Herbicide spraying CDA			5
Insecticide spraying CDA			1
	Totals	17	46
System 3. Using IITA Four-Row Rice Planter Plus Herbicides (No-Till)			
Herbicide (systemic) spraying CDA			5
Planting, using IITA four-row planter			10
Herbicide spraying; preemergent, CDA			5
Insecticide spraying–CDA			1
	Totals	nil	21
Conventional (for Comparison)			
Field preparation using contract tractor for initial tillage		4	120
Seeding (hand dribbling in rows)			300
Manual weed control by hand and hoe, twice			500
	Totals	4	920

NOTES: (1) Yields for each system are comparable; (2) "conventional" data from N.A.F.P.P. surveys.

equipment and man hours. Man hours (and machine hours) compared for alternative systems of upland rice establishment are shown in Table 11-3.

It is imperative that equipment be designed to meet local conditions allowing for the expansion of no-tillage for farmers using many systems of production, whether they be powered by hand, animal or internal combustion engine. Improvement in earlier equipment has been made and growing interest on the part of the grower will accelerate new innovations.

REFERENCES

Wijewardene, Ray. 1979a. Energy and technology in tropical small holder farming. *Congrès International de Génie Rurol.*

Wijewardene, Ray. 1979b. Planters (seeders) for zero-tillage farming systems in the humid tropics. In *Soil Tillage and Crop Production,* International Institute of Tropical Agriculture, Ibadan, Nigeria, *Proc. Series No. 2,* R. Lal, Editor. pp. 161–169.

12
No-Tillage in the Tropics

Grant W. Thomas
Professor of Agronomy
University of Kentucky

and

Robert L. Blevins
Professor of Agronomy
University of Kentucky

and

Shirley H. Phillips
Associate Director of Extension
College of Agriculture, University of Kentucky

In the tropical regions of the world there exists a wide range of differences in climate, vegetation and soil resources. The so-called tropical zone comprises about one-third of the land surface of the earth. It is difficult to clearly define the tropics, and for convenience is often given a latitudinal definition, which includes the zone between Tropic of Capricorn and Tropic of Cancer or latitudes of 23.5° N and S (Kalpagé, 1974). Another definition of the tropics (Sanchez, 1976) is that part of the world where the mean monthly temperature variation is 5°C or less between the average of the three warmest and three coldest months. Located within this tropical zone are many of the developing nations that face food supply problems (U.S. President's Science Advisory Council, 1967) because food production technology has lagged behind population growth.

The general climatic and soil characteristics of the tropics suggest a great potential for food and fiber production. The diversity of soil resources of the tropics ranges from the poorest to the most fertile in the world. Rainfall is the most significant climatic factor that controls food production in the tropics, again ranging from the extreme desertic zones to the tropical rainforests. More research and development are needed before the potential productivity of the tropical areas can be realized. Development of farming systems that are adaptive to and acceptable to native farmers is necessary before there can be a positive impact of modern technology on agricultural production in the tropics. No-tillage or minimum tillage production systems appear to have a potential for many zones within the tropics. However, the technology as developed in the temperate zones of the world must be refined to meet the specific soil, climatic and socioeconomic conditions of the tropics. This will require both imagination and adaptive research efforts to successfully implement this technology.

SOIL RESOURCES

Distribution and Classification of Tropical Soils

When soils of the tropics are mentioned there is a tendency to generalize and think about deep, red, highly weathered Oxisols (sometimes referred to as Laterites and Latosols). Even though these soils are of major importance, numerous other soil orders occur such as Alfisols, Ultisols, Entisols, Vertisols and Inceptisols (Table 12-1). For example, in Africa, the mature, but less weathered Alfisols are equal to Oxisols in hectares. However, in the tropical zone of Latin America, the Oxisols and Ultisols are the major soil orders (Figure 12-1).

In the equatorial regions of the world the effect of climatic factors on soil formation is especially important. The climatic factors have a direct influence on the pedogenetic processes but also control the soil potential for crop production. At lower elevations, less than 2,000 meters, within the tropics, temperature remains high year round, with only a small seasonal variation. Under these isothermic temperature regimes, soil forming processes are accelerated. These high mean annual temperatures also result in rapid break-down of organic matter. At higher elevations of the mountains the mean annual temperature drops rapidly with increasing elevation. As the mean annual temperature drops below 22°C, organic matter often

Table 12-1. Approximate extent of major soil suborders in the tropics (million ha). (From Sanchez, 1976, reproduced with permission of John Wiley & Sons, Inc.)

ORDER	SUBORDER	AFRICA	AMERICA	ASIA	TOTAL AREA	PERCENT
Oxisols	Orthox	370	380	0	750	15.0
	Ustox	180	170	0	350	7.5
		550	550	0	1100	22.5
Aridisols	All	840	50	10	900	18.4
Alfisols	Ustalfs	525	135	100	760	15.4
	Udalfs	25	15	0	40	0.8
		550	150	100	800	16.2
Ultisols	Aquults	0	40	0	40	1.0
	Ustults	15	35	50	100	2.2
	Udults	85	125	200	410	8.2
		100	200	250	550	11.2
Inceptisols	Aquepts	70	145	70	285	6.0
	Tropepts	0	75	40	115	2.3
		70	225	110	400	8.3
Entisols	Psamments	300	90	0	390	8.0
	Aquents	0	10	0	10	0.2
		300	100	0	400	8.2
Vertisols	Usterts	40	0	60	100	2.0
Mollisols	All	0	50	0	50	1.0
"Mountain areas"		0	350	250	600	12.2
Total		2450	1670	780	4900	100.0

accumulates within the soil profiles. Many soil properties of the mountain soils are more similar to soils of the temperate regions.

Properties and Characteristics of Tropical Soils

The contrasting climatic conditions and parent materials result in the formation of soils of widely variable characteristics. This variability is described in the U.S. President's Science Advisory Report (1967) as follows:

> "Tropical soils range from highly-leached ones of the rainforest to alkali-saturated soils of the desert; from rich volcanic soils of Java

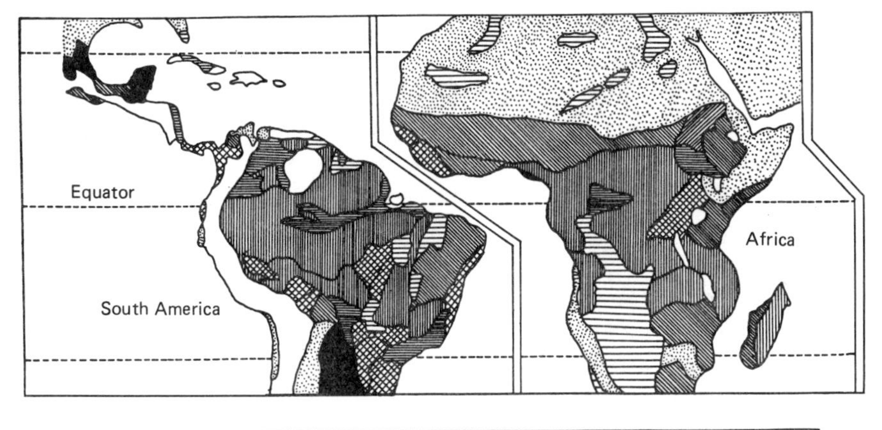

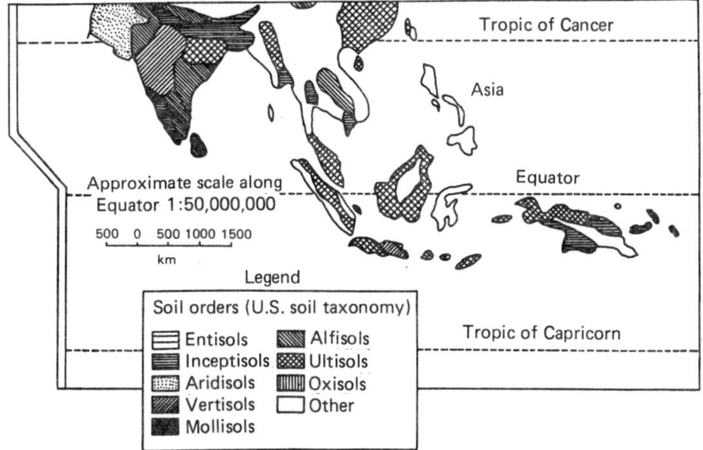

Figure 12-1. Soils of the tropics. (From Kalpáge, 1974)

through alluvial soils of Nile Delta to impoverished soils of the ancient uplands."

In the tropics, age is a factor of soil formation which is expressed through degree and depth of weathering and horizon development. Range land areas in the tropics belong to continental shields and tablelands that have not been subjected to recent folding. However, some gentle upwarping has resulted in continental swells and down-warping into large basins (Aubert and Tavernier, 1972). Because many soils of the tropics have developed in place for long periods of time without disruptions by any cataclysmic upheaval such as glaciers or volcanic eruptions, the soil forming factors have expressed themselves by forming deep, red, highly weathered soils that show strong horizonation, are inherently low in fertility, dominated by kaolinite clay minerals in the clay fraction, retain large quantities of free iron oxides from their parent material (this contributes to the deep red colors of these soils), have low cation exchange capacity, are susceptible to leaching, contain some free aluminum oxides and have a deep acid solum.

Because these highly weathered soils of the humid tropics depend on the recycling of bases from forest or grassland vegetation, they are sometimes referred to as "fragile soils" in which a lowering of organic matter levels may be disastrous. For this reason, when these soils are deforested and brought into cultivation using traditional farming systems the natural soil fertility is rapidly exhausted by leaching. The surface soil soon becomes depleted unless large quantities of residues are returned. However, one must keep in mind that not all of the soils of the tropics are highly weathered. For example, the tropics contain highly fertile alluvial soils along with the younger and often fertile soils of volcanic origin.

According to Buol *et al.* (1973), there are certain misconceptions regarding the classification of tropical soils as Oxisols. Many of these soils do not meet all the criteria of Oxisols. Many deep, dark red soils of the tropics failed to qualify as extremely weathered Oxisols and are more properly classified as Ultisols, or, in some situations, Alfisols and Inceptisols.

The Oxisols and Ultisols are examples of soils of the tropics that have reached an advanced stage of weathering. In the tropical zones the Oxisols usually occupy the older more stable landscape positions. However, not all of the soils of the tropics develop on stable landscapes. Recent mountain building that exposed new rocks to erosion and weathering has occurred in parts of Asia. The mineral supplying capacity of the younger soils developing in these areas is much greater than the highly weathered, infertile soils that characterize the older, more stable landscapes. Another common situation is areas where volcanic ash is the parent material. These volcanic ash soils are young soils (Ando soils) that are presently classified as Inceptisols. These soils have limited profile development and the chemical and physical characteristics are largely determined by the properties of the ash material. Andepts, for example, is the suborder classification of many of the volcanic ash-derived soils that occur in the humid tropics and make up a significant portion of Central America.

In tropical Africa, the younger, relatively high-base status soils (Alfisols) occupy about one-third of the land area. In South America, Oxisols and Ultisols account for major part of the land area, with the remaining made up of Alfisols, Entisols, Inceptisols and some Vertisols (Table 12-1).

During the past 15 years, quite a lot work has been done on the effects of no-tillage versus more conventional forms of tillage on soil chemical and physical properties. Until rather recently nearly all the work involved soils of temperate to subtropical regions. However, the spreading interest in no-tillage has begun to show up in studies on soil properties of Latin America, Africa and Asia. Most of these studies are of short duration thus far, but there are some points that stand out.

The first problem in describing tillage effects on soil properties is the recognition that soils of the tropics are amazingly variable. These soil groups are mostly influenced by parent material and secondarily by temperature and rainfall.

An Andept of Central America may have a bulk density of 0.75 g/cc and an organic matter content of 10 percent. Obviously a tendency of no-tillage to perhaps increase bulk density slightly and increase organic matter percentage would have very little effect on properties of this soil. However, on a Vertisol a slight increase in bulk density might be disastrous. Similarly, a small increase in organic matter on Tropepts might make a very meaningful change for the better, both chemically and physically.

Studies by Lal (1976) on an Alfisol in Nigeria showed more soil organic matter, higher infiltration rates, and more extractable Ca, Mg and K in no-tilled than in plowed soils.

In the humid and subhumid tropics the changing of land use from forested to an arable land use often results in a rapid deterioration of soil physical properties. Lal *et al.* (1980) in Nigeria on an Oxic Paleustalf conducted studies comparing differing rates of surface mulch. Increased levels of mulch resulted in higher total soil porosity, increased field infiltration rates, higher saturated hydraulic conductivity and greater moisture retention. The chemical fertility of the soil decreased with time after forest removal. However, the rate of degradation was less as the mulch rate increased. The continuous ground cover afforded by a mulch in this study reduces erosion losses, prevents soil crusting and suggests that these soils may be used for food and fiber production without soil quality losses and reduced productivity. Lal *et al.* (1980) suggest that a practical way of obtaining and maintaining a surface mulch on these fragile soils is through a no-tillage farming system. This system would utilize residues from the previous crops and dead weeds to provide a surface mulch.

The main advantage of no-tillage cropping of soybeans with wheat as observed by Hayward *et al.* (1980) under tropical conditions is the reduction in soil erosion losses. The mulch of this system reduces the rate of degradation of soil aggregates and the mulch plus the root systems favors the maintenance of a higher level of organic matter as compared to a cultivation system.

Soil Fertility

Since there exists a great diversity in soil properties in the tropics a variety of soil fertility problems and needs would be anticipated. Some of the more basic soil fertility situations will be discussed below and how they may be influenced by no-tillage farming. Because of the infancy of no-tillage farming systems little research data is available from the tropics.

Many soils of the humid tropics are acid with soil pH values less than 6. Higher pH values are observed in the more arid regions of the tropics. The Ultisols are especially acid with exchangeable aluminum causing the acid infertility problem. The acid infertility reduces yields by causing either aluminum toxicity, calcium or magnesium deficiency, manganese toxicity or a combination of these factors. The Oxisols are characterized by low levels of nutrients, low exchangeable acidity and low levels of exchangeable aluminum. These soils should not be limed using the same approach used in the temperate region. According to Sanchez (1976), liming in the tropics should be approached with the application of enough to neutralize the exchangeable aluminum and this can usually be accomplished by raising the soil pH to 5.5. If manganese toxicity is a problem, it may be advantageous to lime to pH 6.0.

Overliming can create numerous problems, such as inducing micronutrient deficiencies, reducing available soil phosphorus and resulting in decreased soil aggregate stability. Selection of crops or cultivars within crops that have a high aluminum tolerance is one practical and very popular approach to dealing with the inherent soil acidity. Liming programs in the tropics should be based on chemical tests to ascertain the minimum level of lime needed. Minimum levels of applied lime plus use of deep rooted crops, and acid-tolerant varieties are management methods that have potential.

The nutrient most commonly causing reduced yields in the tropics is nitrogen. Most soils in the tropics are deficient in nitrogen except for newly cleared land. This is another reason for the popularity of the shifting cultivation or slash and burn type of farming commonly used in the tropics. In the more humid tropics, large rates of dry matter are returned to the soil, often five times more than in the temperate regions. However, the higher year round temperatures and moisture provide an environment where organic matter decomposes about five times faster also. The net effect is that tropical soil usually contains about the same amount of organic matter as temperate soils. The problem develops when the equilibrium is disturbed by reducing plant residues returned to soil each year. Nitrogen fertilizer recommendations must be related to crop requirements and soil properties. Field experimentation is probably the best and most useful way to evaluate crop and soil responses to added N fertilizer.

Most Oxisols and Ultisols in the tropics are deficient in phosphorus. This supports the theory that tropical soils usually show declining levels of total phosphorus with increasing weathering (Olson and Englestadt, 1972). The rate of loss of phosphorus also depends on cropping intensity and erosion losses. The level of phosphorus deficiency can be effectively predicted by soil test. Supplying phosphorus fertilizer to correct deficiencies often requires careful management. For example, tropical soils with high P-fixing potential may require banding applications of P fertilizer. Liming to adjust the soil pH to 5.5 may be helpful by precipitating exchangeable aluminum so the aluminum will not fix the added phosphorus. The few soils of the tropics that may have sufficient levels of P would be younger alluvial soils.

Our experiences on Inceptisols and Vertisols in the Dominican Republic using no-tillage methods to grow corn and soybeans indicate that these soils are highly responsive to both nitrogen and phosphorus fertilizer amendments. Lal (1981) reports greater fertilizer use efficiency for no-tillage maize production in Nigeria than with conventional tillage.

The effect of cultivation of tropical soils after clearing, especially the Ultisols, is described by Baver (1972) as follows: First, the soil is unprotected from the impact of raindrops and the intense heat rays of the sun; there is an increased level of activity of microorganisms which increases the rate of organic matter decomposition; soil aggregates become less stable, resulting in slower infiltration, which increases water runoff and soil erosion losses. According to Baver, these changes may take place rapidly, on the more acid Ultisols, even within one to two years of cultivation. The better structured Oxisols and Andepts (volcanic ash soils) are less susceptible to this rapid deterioration of physical properties resulting from cultivation. Use of a shifting cultivation system of farming is often used in the tropics to avoid irreversible damage to the more fragile soils. Continuous cropping may be an alternative if minimum or no-tillage is used, mulches maintained at the surface and fertilizer and lime used.

Organic matter maintenance, improvement or reduced decomposition is highly important in all crop production areas of the world. In the tropics the maintenance of present organic matter levels in the soil is a major problem. No-tillage can improve this situation by actually increasing the organic matter levels. Roman and Barker (1978), Rio Grande do Sul, Brazil, reported increased organic matter percentage after several consecutive maize crops.

As would be expected, the organic matter under no-tillage was much higher in the 0–5-cm depth. There was a trend to higher organic matter content in the deeper zones of 5–15 cm in the soil profile. Part of the increase could probably be identified with increased vegetative residue associated with higher yields. Thus increased organic matter would have a positive effect on reduced soil temperatures, increased rainfall infiltration and more productive soils.

Numerous researchers have reported that mulches on tropical soils improve soil chemical and physical properties (Juo and Lal, 1977; Lal, 1975; Meyer, 1975; Muzilli, 1981; and Manipura, 1972). According to Lal *et al.* (1980), the gradual replacement of forest canopy by a suitable crop residue mulch will slow the normal rate of decreasing productivity that usually accompanies clearing the land and that with good management soil productivity can be maintained at a suitable level. No-tillage or reduced tillage systems that include maintaining mulches at the soil surface may be a viable alternative system to use in the tropics according to Lal *et al.* (1980).

Erosion Problems

The poor selection of land use patterns is responsible for much of the erosion in the tropics. The very steep slopes should be left in permanent cover. Cultivation up and down the slope and excessive tillage operations can also add to the erosion losses. Soil water erosion is often greater in a climatic zone that has a dry-wet season than in regions with a more equal distribution throughout the year. The high intensity rainstorms during the rainy season are very erosive. At the end of the dry season the soils are often crusted and vegetation cover is sparse. These conditions make the soils especially susceptible to erosion losses when the first rains occur.

Because of the need for more arable land to produce food and fiber it seems inevitable that large areas of tropical soils will continue to be deforested and used for row crop production. The major consequence of introducing continuous cultivation (Lal *et al.*, 1980) is the progressive deterioration of soil structure. This destruction results in crusting at the soil surface and subsequent reduced infiltration rate. This brings about a situation that favors high rates of soil erosion (Goswami, 1971). The high-energy rainstorms of the tropics are highly erosive when soil structure deteriorates. The decrease in aggregate stability may also be intensified by a declining level of organic matter that usually occurs in tropical soils under intensive cultivation (Drosdoff, 1972).

Mulch or no-tillage farming systems offer an improved system of land use for these fragile tropical soils which are highly erodible and difficult to manage. Long term field trials by Lal (1976) showed benefits from no-tillage through reduced soil temperatures and increased soil moisture as compared to plowed and bare fallow plots. When no-tillage was used for maize production, soil water runoff and soil erosion losses were dramatically reduced (Table 12-2).

Considerable research efforts have been made in Brazil to deal with the erosion problems on tropical soils when cultivated and used for soybean production. In Paraná, Brazil (Table 12-3), under natural rainfall and with a soybean-wheat double crop system, soil

Table 12-2. The effect of tillage systems on soil water runoff and soil erosion under maize for a rainfall occurrence of 44 mm. (From Hayward *et al.*, 1980, by permission of *Outlook on Agriculture.*)

TILLAGE	SLOPE	RUNOFF	SOIL LOSS
	(%)	(%)	(mt/ha)
No-tillage	1	1.2	0.0007
	5	1.8	0.0007
	10	2.1	0.0047
	15	2.2	0.0015
Ploughed	1	8.3	0.04
	5	8.8	2.16
	10	9.2	0.39
	15	13.3	3.92
Bare fallow	1	18.8	0.2
	5	20.2	3.6
	10	17.5	12.5
	15	21.5	16.0

Table 12-3. Soil losses from three different soils under natural rainfall – Paraná, Brazil. (From Hayward *et al.*, 1980, by permission of *Outlook on Agriculture.*)

	SOIL LOSS (mt/ha)		
TILLAGE	HAPLUDULT	HAPLORTHOX	HAPLOHUMOX
No-tillage	0.09	1.9	0.6
Conventional	2.4	2.3	4.3
Bare soil (w/o crops)	14.2	4.9	35.5

losses were drastically reduced when compared to a bare soil (Hayward *et al.*, 1980).

Soil erosion studies and surface water runoff measurements were made in the semi-deciduous rainforest sector of Ghana (Mensah-Bonsu and Obeng, 1979). Surface mulch treatment produced the lowest soil erosion loss (Table 12-4) of the cultural practices evaluated. Corn grain yields were highest on the mulched followed by the no-tillage. Erosion losses on the no-tillage during the initial year of the study were four-fold greater than in years two and three because there was no surface residue at the beginning of the experiment.

In many areas where slope and rainfall patterns combine to produce serious erosion problems, multicropping can reduce soil losses by maintaining crop residues at the soil surface to retard runoff. Maintaining a crop canopy and growing plants can effectively reduce dislodging of soil particles by breaking the impact of rain droplets on unprotected soil.

WATER RESOURCES

Since the growing season in the tropics is not controlled by freezing temperatures, the availability of water becomes the controlling influence of the cropping sequence and date of planting. Soil and

Table 12-4. Runoff and soil loss on a 7.5 percent slope in Kwadaso – Kumasi, Ghana; average of three years 1974–76. (Reprinted from *Soil Physical Properties and Crop Production in the Tropics,* Mensah-Bonsu and Obeng, in Lal and Greenland [Editors], copyright 1979, by permission of John Wiley & Sons, Ltd.)

TREATMENT	RUNOFF	SOIL LOSS
	(% of rainfall)	(mt/ha/yr)
Bare fallow	43.9	200.0
Zero-tillage	6.9	3.8
Mulching	1.9	0.4
Ridging (across slope)	7.4	3.6
Minimum tillage	8.8	3.2
Mixed cropping	13.5	18.8

crop management planning must be synchronized with the beginning and ending of the rainy season. Thus, the potential productivity of tropical soils is closely correlated with moisture supplying potential and soil characteristics.

Using a climatic definition for identifying the tropics based on temperature, regions that have occasional frosts or have a mean monthly temperature less than 13°C are often excluded (U.S. President's Science Advisory Report, 1967). Thus the classification of tropical climatic zones are usually based on the amount and seasonal distribution of rainfall.

Climatic Conditions in the Tropics

Soil temperature has less influence on cropping patterns and tillage systems in the tropics as compared to the temperate regions. For example, in the tropics the mean monthly temperature varies less than 6°C for hottest versus coolest months of the year with the least temperature variation occurring near the equator. The highest mean annual temperature occurs in the desert areas near the Tropic of Cancer (Sanchez, 1976). The annual distribution of monthly mean temperatures is presented in Figure 12-2, for several locations.

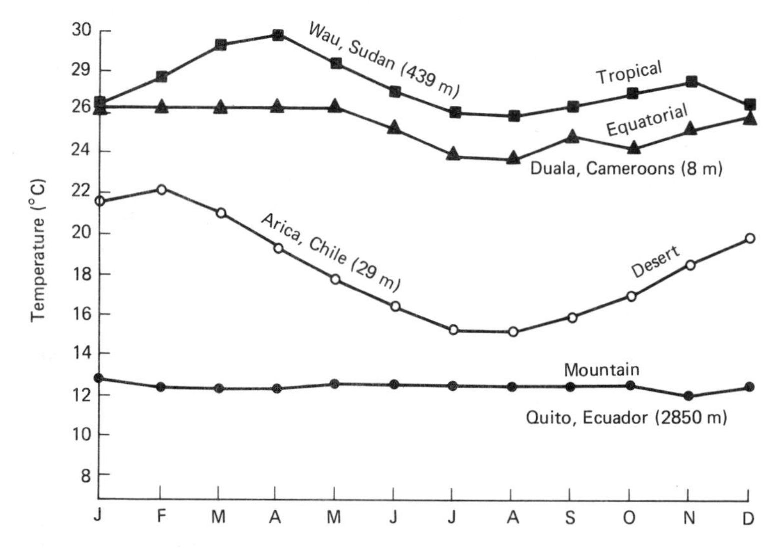

Figure 12-2. Annual temperature distribution at selected stations in the tropics. (From Kendrew, 1961)

Note the uniformity of the annual temperature recorded at Quito, Ecuador near the equator. The cool but constant temperature recorded at Quito is associated with the high elevation 2,850 m above sea level.

Solar radiation is usually higher in the tropics than in temperate regions. A higher percentage of the solar radiation received in the tropics is also available for photosynthesis on a yearly basis because temperatures are above the minimum required for plant growth throughout the year. The highest solar radiation values are received on the fringes of the tropics in more arid regions. For example, Sudan has recorded an average of 600 langleys/day. In contrast the lowest figures for the tropics usually occur for portions of the Amazon and Congo rainforests due to cloud cover.

Rainfall as mentioned before is by far the most important climatic factor affecting tropical agriculture. The variation in rainfall, for example, ranges from zero to 10,000 mm in the tropics. These include the equatorial or rainy climates to the aridic tropical deserts (Figure 12-3). The dry climates that include the tropical deserts makes up about one-fourth of the land area of the tropics. This semi-arid to arid region has less potential for using no-tillage farming. Problems of acquiring a surface mulch and lack of adequate rainfall to produce a crop are severely limiting factors.

Monthly rainfall-evapotranspiration balances at selected tropical locations are included in Figure 12-4. The different rainfall climates are represented along with locations representing Tropical America, Africa and Asia.

Despite very high total rainfalls, drought in the seasonal tropics is a regularly occurring problem. Since these tropical areas have distinct wet-dry seasons, part of the year is not very productive for plant growth despite the favorable temperatures. In addition, whether or not there are wet-dry cycles to contend with, there is great variability in rain so that there may be exceedingly dry periods at unexpected times.

Soil Moisture Storage and Water Use

Crops grown using no-tillage systems that maintain a mulch at the soil surface are less subject to drought stress than those grown with clean-tilled production systems (Table 12-5). The use of the surface

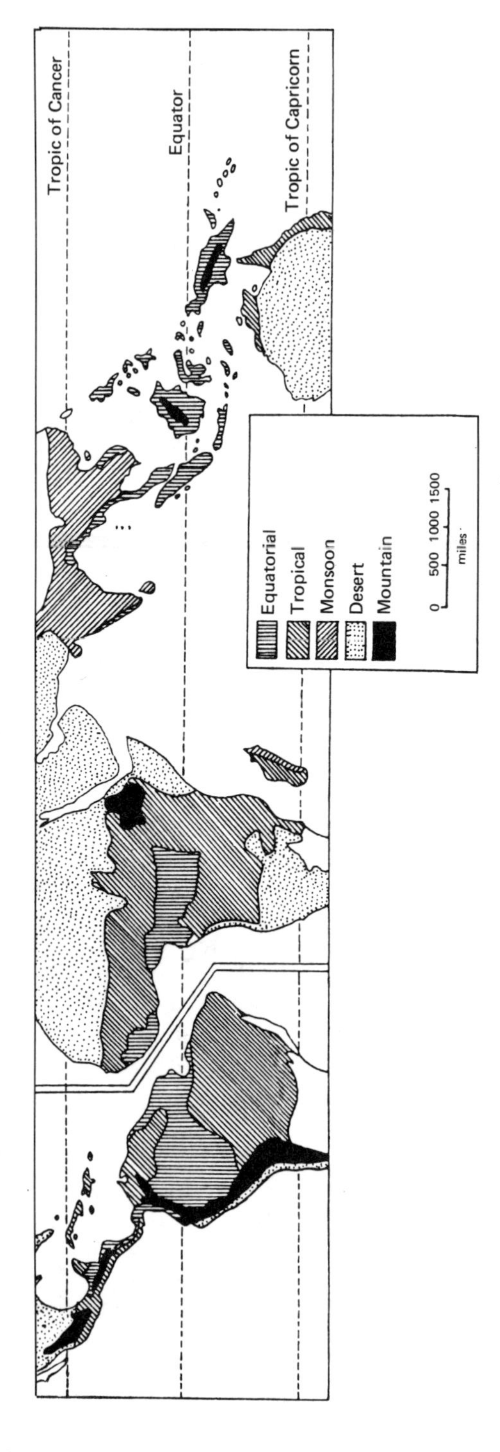

Figure 12-3. Climatic regions of the tropics. (From Kalpagé, 1974)

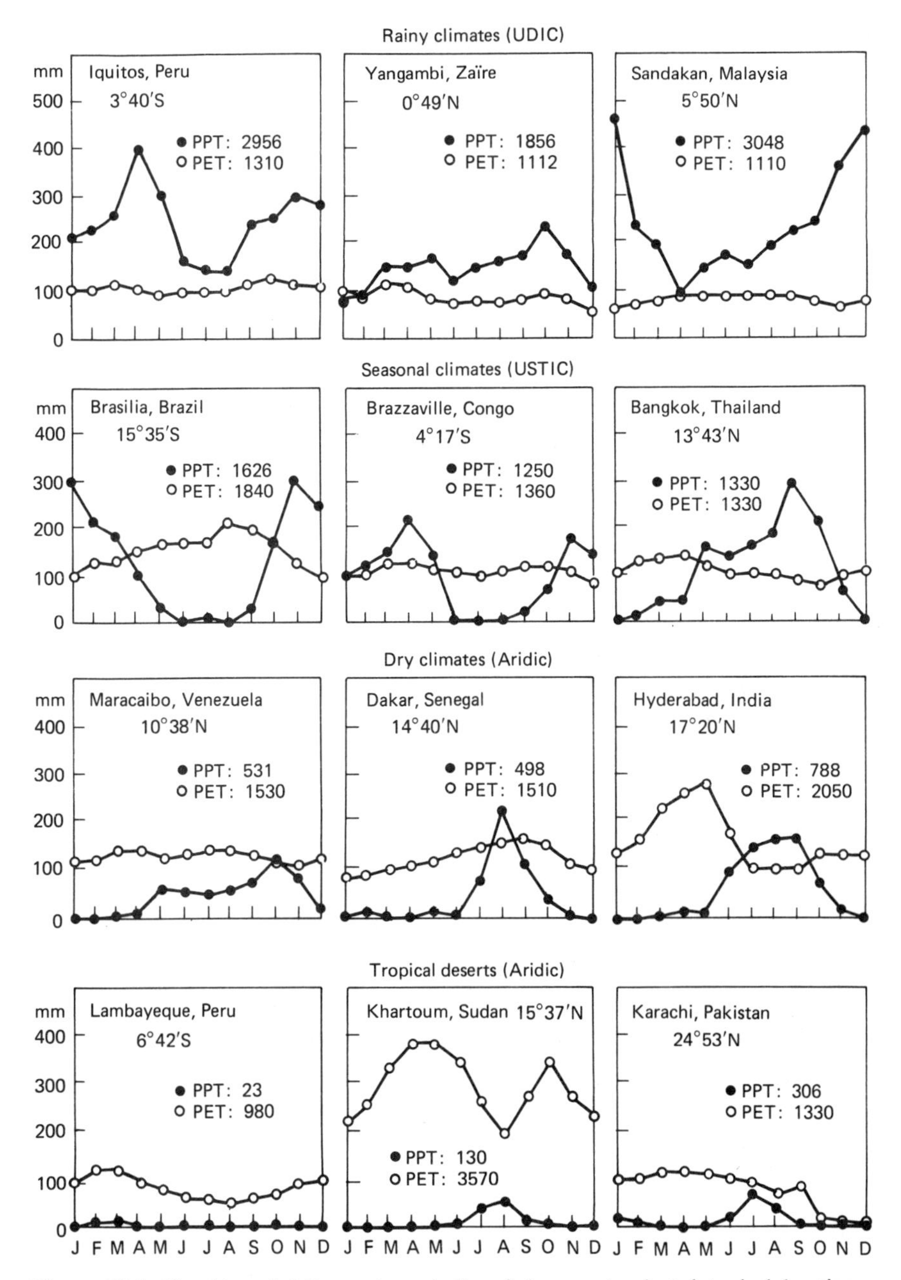

Figure 12-4. Monthly rainfall-evapotranspiration balances at selected topical locations. PPT = precipitation, PET = potential evapotranspiration. Annual totals in numbers. (From Sanchez, 1976)

Table 12-5. Effect of tillage systems on soil moisture retention two weeks after planting various crops. (From Hayward *et al.*, 1980, by permission of *Outlook on Agriculture.*)

	PERCENT MOISTURE RETENTION, 0–10 cm DEPTH			
TILLAGE SYSTEM	MAIZE	PIGEON PEA	SOYA	COWPEA
No-tillage	13.3	12.1	10.6	15.4
Plowed	9.7	10.8	7.3	12.3

mulch not only reduces runoff but reduces the rate of soil water evaporation during the first and second stage of soil water evaporation. This increases the duration of plant available water by several days depending on the soil characteristics. For a short-term drought, no-tillage is always more favorable than conventionally tilled treatments.

No-till plots in western Nigeria (Lal, 1981; and Lal *et al.*, 1982) had four to five times more production of earthworm casts. Both soil structure and soil porosity was improved by this mixing of mineral and organic components of the soil. The improved soil structure and thoroughly mixed organic matter improved the moisture retention capacity of the soils.

Improved moisture retention by using no-tillage in the tropics is critical on the drought-susceptible, shallow soils. Soils that are low in organic matter content and subsequently exhibit poor soil structure commonly lose up to 50 percent of the rainfall by surface runoff. Consequently, improved infiltration rates associated with and undisturbed, surface mulched top-soil is a key factor in conservation of soil water. Obviously this would not be a factor on certain Inceptisols of volcanic origin that have high organic matter and low bulk densities to begin with. By using no-tillage systems, cultural practices and yields can be sustained for a longer period than with conventionally tilled soil.

The soil moisture regimes common to the tropics have been identified as udic, ustic, aridic and aquic by Sanchez (1976). The classification uses the terminology and definition of U.S. Soil Taxonomy System (Soil Survey Staff, 1975). The udic soil moisture regime implies that for most of the year water stress will not be a factor. This regime corresponds to the rainy or equatorial climate of the tropics and represents about 29 percent of the tropics. The ustic moisture regime describes soil conditions that have a definite dry season of several months and relates to the seasonal or monsoonal climates. This regime includes about one-third of the soils of the tropics. The aridic implies an even longer dry season than the ustic and relates to the dry desert climate. About 29 percent of the land area is aridic. Aquic moisture regimes are associated with poorly drained sites.

According to Lal (1981), climatic conditions that favor the use of no-tillage farming system include:

1. Regions characterized by high intensity rainstorms with thunder showers in the beginning of the rainy season after a prolonged dry period.
2. Regions where there are two or three periods of seven to ten rainless days during the rainy season.
3. Regions where the precipitation is unpredictable and total annual rainfall exceeds 500 mm.

There appear to be definite advantages to growing millet and sorghum using a no-tillage or reduced tillage system in the semi-arid regions of the tropics. In the more humid, rainy and seasonal climates, crops such as corn, soybeans, cowpeas and pigeon peas can be successfully grown using no-tillage.

Conservation of Water Using No-Tillage

The use of no-tillage has advantages under tropical conditions for a number of reasons. First, the slower rate of evaporation of water from a mulched soil surface can extend the time that the plant can grow normally under dry conditions just as it can in more temperate regions. Second, a higher proportion of the high-intensity rainfall, which is common in the tropics, will be saved under no-tillage. Runoff is lower because the water infiltration rate is higher where there is a surface mulch and where natural soil structure is maintained. Third, soil erosion is reduced significantly so that the surface soil is preserved, giving a naturally higher available water content for plant growth.

Depending on the sunlight intensity, soil temperature will be reduced somewhat with a surface mulch (Marelli *et al.*, 1981). This can be an advantage under some conditions and with certain crops. However, it should be emphasized that air temperature just above the soil will usually be higher than with conventional tillage so that young seedlings that are heat-sensitive may suffer more with a surface mulch than without it.

HUMAN RESOURCES

Availability of Labor

In addition to the advantages of water conservation by using no-tillage that have been cited above, there is another advantage that, in the authors' opinion, is even more decisive – the saving of human energy. On every farm in the world, there are periods of time when labor shortages occur. One of these times occurs when land is being prepared for crop planting. There is an almost desperate haste to ready the land and, as likely as not, this effort will be delayed by wet and/or dry weather.

No-tillage can reduce the labor bottleneck for tropical farmers even more than it does for farmers in the temperate zone. It is a common myth that, in the tropics, where laborers' wage scales are apt to be a fraction of those in more developed countries, it is somehow profitable to have field laborers busy at non-productive jobs. The feeling is that with very low labor costs, there is more economic benefit to be gained by using a man for weed control than a herbicide. This idea is incorrect on several counts. First, and perhaps most important, there is always needed work to be done on a farm, whether it be tropical or temperate. Much of this work has more to do with making a profit than the dull drudgery of hoeing, assuming that weeds could be controlled in other ways. Second, just because laborers are paid badly does not mean that they enjoy endless, thankless work any more than higher paid employees, or that they perform any better. Third, the cost of paying many laborers a small wage is still very high, in fact oppressively high unless the crop is of high value. An illustration may suffice: In the Dominican Republic, we paid $18 to have 0.06 ha prepared by hoe hands with a $5.00 daily wage. An area the same size and adjacent was treated with herbicides for $3.00, and the $3.00 treatment included season-long weed protection whereas the $18.00 did not. Hence, the balance in favor of herbicides is even greater under small-farm, tropical conditions than it is in more developed agriculture.

It may be argued that in some cases only family labor is involved and that no money is exchanged. Such cases are fairly rare. Even the smallest farmers in the tropics tend to hire help even if it is only to keep up appearances. In fact, they are much more likely to hire help than are farmers in the more developed countries. Even if they do not hire help, there usually is something more profitable that they can be doing with their time than hoeing.

Another factor closely related to the cost of labor is the timeliness of operations. This usually is tied closely to availability of water, either too much or too little. If there is too much water and a rainy season is approaching, planting of the crop may be delayed or as too often happens, completely missed. With no-tillage, on the other hand, spraying and planting may be accomplished with only small breaks in the weather. When the weather is too dry, cultivation may either be delayed or the soil cultivation may cause the loss of most soil water. In either case, emergence of the crop will be delayed and crop and financial losses are likely.

It has been our experience in Kentucky that while farmers often list erosion control as a primary reason for using no-tillage, a more compelling reason, for successful no-tillage farmers, is the flexibility it gives in terms of time, soil water content and weather. The same factors operate in the tropics as well. Hence, a good control of erosion may be achieved but the true motivation may well have been the desire to save time and trouble.

The displacement of farm laborers by no-tillage (or by some other labor-saving practice) has not been studied to any great extent under tropical conditions. It is probable that on large farms or plantations some labor will be displaced. On small farms, however, the tendency probably will be to increase the intensity of operations or the area farmed. The net effect should be an increase in productivity per man as well as an increase in productivity per hectare of land. The effect on prices of commodities will probably be bad for the farmer as it generally is under new technology.

Other Sources of Energy

In addition to human labor, the use of fossil fuel is increasing as agricultural production is augmented in the tropics. The principal uses of petroleum or its substitutes are found in fertilizers, chemicals and in tractor fuel. No-tillage does not eliminate the need for these energy sources, but it does reduce the total required rather substantially. Table 12-6 shows the diesel fuel used for upland rice in Brazil (de Andrade, 1982) using conventional tillage and no-tillage. Note that with upland rice grown conventionally 11 trips were made over the field with a total of 111 liters of diesel fuel used. With no-tillage only 4 trips and 48 liters of fuel were used.

If, instead of tractor fuel, one substitutes animal power, this is not without cost either. Just the keeping of an animal involves land cost which could be devoted to edible and/or saleable crops. It is often more economical, quicker and easier to control weeds with a hand sprayer than to keep, hitch up and unhitch an ox or water buffalo. Given the mix of crops and the small areas devoted to any one crop, the use of herbicides and hand sprayers to eliminate tillage can be the most economical single practice undertaken. Of course, the key to making it economic is to at least keep crop yields equiva-

Table 12-6. Trips over the field and fuel consumption of conventional and no-tillage systems for irrigated upland rice. (From de Andrade, 1982.)

OPERATION	CONVENTIONAL TILLAGE		NO-TILLAGE	
	NUMBER OF TRIPS	FUEL USED	NUMBER OF TRIPS	FUEL USED
		1/ha		1/ha
Plowing	1	10	0	0
Discing	5 or more	45	0	0
Fertilizing	1	9	0	0
Planting	1	9	1	10
Seeding cover	1	9	0	0
Herbicides	1	9	2	18
Harvesting	1	20	1	20
Total	11	111	4	48

lent to those obtained under conventional tillage. This does not seem to be a problem under tropical conditions (Tables 12-7 through 12-9). In fact, in many cases yields are increased greatly (Huntington *et al.*, 1982; Muzilli and Ique, 1979; Pérez, 1982; Shenk, 1979; and Zaffaroni *et al.*, 1979). If yield increases, reduction in energy use, timeliness and (perhaps most important) increased income are dependable results of no-tillage in the tropics, the system has a bright future there.

Acceptance of New Technology

The tropics have a wide range in agricultural technology. For plantation crops, such as banana, for example, every chemical and cultural treatment known is implemented immediately. Banana, therefore, is a classic case of direct technology transfer, without any delay at all. With crops grown on small farms, the technology is delayed

Table 12-7. Effect of tillage on yield and net return from corn in the Dominican Republic. (From Pérez, 1982.)

TYPE OF TILLAGE	YIELD	NET RETURN
	(kg/ha)	($/ha[1])
Conventional	4,420	363
Reduced	4,110	343
Minimum	4,820	488
No-tillage	4,790	560

[1] Republica Dominicana dollars.

Table 12-8. Red bean yields in Corrocitos, Dominican Republic as influenced by fertility and tillage. (From Huntington *et al.*, 1982.)

	YIELD OF RED BEANS (kg/ha)	
FERTILITY TREATMENT	CONVENTIONAL TILLAGE	NO-TILLAGE
Check	42	70
100 kg N, 75 kg P_2O_5	194	377

Table 12-9. Yield of corn alone and of corn and beans intercropped and net income at Turrialba, Costa Rica as influenced by tillage. (From Zaffaroni *et al.*, 1979.)

TREATMENT	YIELD CORN ALONE	NET INCOME	YIELD OF INTERCROPPED CORN	YIELD OF INTERCROPPED BEANS	NET INCOME
	(kg/ha)	($/ha)	(kg/ha)	(kg/ha)	($/ha)
Weeds chopped at the ground, sprayed with roundup	3,389	393	2,485	522	506
Weeds chopped at 50 cm, sprayed with roundup	2,955	310	2,438	458	459
Weeds chopped at ground	505	-89	570	40	-68
Plowed, no herbicide	2,094	114	1,366	403	204
Plowed, sprayed with roundup	2,480	132	1,443	487	192

somewhat, depending on the emphasis or, perhaps, the economic importance of the crop. With rice, for example, technology transfer has been quite good and rice yields in many parts of the developing world have soared. With corn, on the other hand, only spotty acceptance of improved varieties, fertilization and weed control is evident at present.

The above stated situation implies that technology will be accepted roughly in accordance with its effect on financial return to farmers. This further implies that no-tillage technology will be accepted only if it makes a sizeable difference in farmer income. Otherwise, it will remain an interesting but ineffective package of information. If this reasoning is correct, then it is important that no-tillage be tried on the crops most important to farmers in a given location.

In the Dominican Republic, Perez (1982) using corn, found that no-tillage gave a net income of $560 compared to $363 for conventional tillage, a difference of $197 per hectare (these values are Republica Dominicana dollars) with only a small difference in yield in favor of no-tillage (Table 12-7). Substantial economic differences such as this can be very instrumental in speeding the technology of no-tillage from the research institute to the farm.

If our experience in Kentucky relates at all to the tropics, and there is no indication that it does not, then the real spreading of a practice is done by the farmers themselves. However, it tends to spread among farmers of about the same type (farm size and income) rather than from large farmers to small farmers. For example, technology used on a plantation does not generally influence small farms which border the plantation. Some of the small farmers will work on the plantation but the ideas do not transmit very well from a grand operation to a small farm unit.

Instead, farmers of a certain class adopt what they see that is good and appropriate from neighbors who are in about the same economic situation as they are. Operations of affluent farmers are not impressive to small farmers at all since it is supposed that "you can do anything with that kind of money."

Once the use of no-tillage is shown to be practical by one farmer, his neighbors will be quick to try it also. Some will make mistakes, be disappointed and vow never to get "fooled" again. Others will gradually make the practice work satisfactorily. Soon, the practice will be well-established in the neighborhood and the innovative farmers will come to that neighborhood to see how it works. In not too much time *other* neighborhoods will be confidently following the practice. The advantage of this means of spreading the technology of no-tillage is that as new people try the system, they will receive support and advice from their neighbors. Soon the entire neighborhood has a vested interest in the success of no-tillage, as if a village club had been formed.

It appears that the best way to get a farmer started on any practice is to place a demonstration or experiment on his farm and then to use it to teach him proper procedures. If the demonstration shows success the farmer may become a disciple. If it is less than successful, his attitude may range from skeptical to sympathetic, but seldom hostile. In addition to observing and learning, the farmer will be a host to visitors. As such he will have a personal interest in the success of the experiment on demonstration. Once he has that attitude, he will be an active participant and this is the most valuable part of the demonstration.

Acceptance of new technology is really a matter of finding some key spreaders at several economic and social levels. These people should be innovative but steady farmers who have good reputations in their community. Finding them may be the most important part of spreading the no-tillage technology.

Management Ability

It often has been pointed out that no-tillage takes better management than conventional tillage. This usually is true because with mechanical tillage, rescue operations such as cultivation for weeds are easy to achieve. With no-tillage, unexpected weed problems require: (1) choice of a herbicide, (2) timeliness and (3) sometimes protection of the crop plant by shielding. In addition to possible weed problems with no-tillage there is the problem of planting seed in unprepared soil. This requires judgment as to soil water content, proper depth of planting and soil temperature. Proper spraying of existing vegetation is also important because strips of live vegetation can greatly reduce crop yields.

A specific management requirement for no-tillage is knowledge of herbicides, their attributes and weaknesses for any weed problem likely to arise. It usually is observed that weed problems change with no-tillage, and that perennial weeds become bigger problems with time. In some cases the perennial weed problems become so severe that no-tillage must be abandoned for a time.

However, even though better management is required for no-tillage farming, the requirements are not impossibly high. The major requirement is that the farmer be timely in his operations and in his observations. Many problems that arise with no-tillage are reasonably easy to correct if they are handled in a few days. Let go for two weeks, the problem may well ensure loss of the crop. Timeliness probably has more to do with successful farming than any other attribute. It certainly is important in no-tillage farming.

PROSPECTS FOR SUCCESS OF NO-TILLAGE IN THE TROPICS

There exists a sound basis for no-tillage in the tropics: with perennial crops such as coffee, cacao and banana, it already is being practiced. In general, with all these crops, shade and a natural residue mulch accomplish most of the weed and grass control required. Where grasses and weeds continue to flourish because of missing plants, ditches or waste areas, the use of a non-selective herbicide such as paraquat or roundup is a routine procedure. Therefore, there is already a strong precedent existing for the non-disturbance of soils, the use of a natural mulch and application of herbicides.

The availability of herbicides, insecticides and fertilizers throughout the tropics is fairly good, and prices are relatively much lower than prices for machinery. Further, the use of small sprayers is so general that they are ubiquitous in most countries. Planting of row crops by hand, machete, stick or hand planter is accepted so that essentially all the requirements for no-tillage to succeed are already in place.

Another recent development is likely to advance the popularity of no-tillage in the tropics. After using corn in a most inefficient way for many years, the tropical countries have at last realized that given good growing conditions, corn can provide what North Americans have known for a long time, the most efficient conversion of sunlight into high-energy food. As a result, then production of corn shows promise of becoming a booming business in many parts of the tropics. For example, the winner of the Phillippine corn growing contest averaged over 10 tons/ha in 1981 and yields of 8 tons/ha are common. When this is contrasted with the pitiful 1–2-ton yields that are obtained with no fertilizer, poor grass control and mediocre varieties, it is clear that these outstanding farmers are incorporating varieties, fertilization, weed and insect control as a package to achieve these superior yields. These are the farmers who will be most receptive to no-tillage or reduced tillage and who will benefit the most from it. Growers with a one-ton average yield will not.

Side by side with corn comes the soybean, so that corn-soybean cropping sequences are becoming more common. In the province of Los Rios in Ecuador, for example, banana and cacao plantations have been replaced by corn-soybeans (two crops per year) so that the entire area looks like a mini-cornbelt.

When subsistence farms or plantation crops are replaced by corn or corn-soybeans, erosion begins to be a serious problem immediately. Producers are vaguely aware of the problem at first and then acutely aware after several hard rainstorms remove a sizeable proportion of the surface soil. Some growers realize that the longevity of their farming base is going to be quite short if something is not done. Others have the attitude that the total soil is quite deep but are aware that productivity is not as high when the subsoil is farmed.

When this realization sinks in, farmers are often receptive to changes in farming practice that *do not reduce their net income.* In contrast to what might be believed, these farmers are among the most receptive in the world to new ideas. In the space of two to five years they have gone from farmers having no experience with corn and soybeans to expert producers who are very familiar with varietal characteristics, populations, appropriate fertilizers, and agricultural chemicals for optimum production. It is relatively easy to get them to try no-tillage when everything else is so new. All that remains is the establishment of enough experiments and demonstrations to convince them that it is worth their while.

With several of the existing cropping systems, no-tillage may be difficult to achieve. Intercropping rows of cassava with beans and sweet potatoes, for example, makes complete reliance on herbicides difficult. Cassava requires a long growing period so that intercropping with short season crop is a useful way to go but it will require more innovative methods of no-tillage than are now available.

It is most apparent from the literature and from many observations that crop production schemes be considered for each of these major production systems in the tropical areas. The opportunity to increase the acreage and intensity of multi-cropping without the risk of reduced yields offers opportunities not available in the temperate zones. Those advantages listed in Chapter 1 for no-tillage are multiplied with respect to tropical zones. The no-tillage systems will become an integral part of improved cropping systems for improved soil management.

Multicropping potential is unlimited in tropical zones compared to temperate areas. One of the most intensive multicropping systems on a commercial basis was observed in Argentina. This system of producing a crop of soybeans, grain sorghum and wheat in a 12-month period offers excellent opportunity where climatical factors are favorable. This system is not without pressure on management skills such as timing of crop establishment and harvest, selection of pesticides compatible with three species following each other at short intervals, increased risk associated with rainfall patterns, mechanization requirements and labor demands.

One of the most popular crop rotations used in the world is a double cropping system combining soybeans and wheat. Researchers in Brazil (Hayward, *et al.*, 1980) compared tillage systems and the yields of soybeans and wheat grown on the same soil each year of the crop rotation (Table 12-10).

Table 12-10. Soybean and wheat yields in a double crop rotation in Brazil. (From Hayward *et al.*, 1980, by permission of *Outlook on Agriculture.*)

TILLAGE SYSTEM	LOCATION	SOY-BEANS 1977	WHEAT 1977	SOY-BEANS 1978	WHEAT 1978
			(mt/ha)		
No-tillage	(1) Rio Grande do Sul	2.6	1.1	3.0	–
	(2) Paraná	3.2	1.0	1.4	2.2
Conventional	(1) Rio Grande do Sul	2.7	0.9	2.8	–
	(2) Paraná	3.1	0.7	0.6	0.6

These data support an increase in wheat yield and soybean yield using no-tillage techniques, with significant increased returns and economic advantage for adoption of no-tillage double cropping.

Small Farmers and Large-Scale Farmers

As stated previously, the tropics are a mix of farm size, ranging from huge plantations to a few tenths of a hectare (Greenland, 1975). The adoption of no-tillage by any size farm is feasible. However, the equipment used for a no-tillage system will be quite different (Wijewardene, 1980).

On a very small farm, needed equipment will be a hand sprayer, a machete and perhaps a hand planter and a shield for the sprayer nozzle. As one moves to a somewhat larger farm, the one addition might be an animal-drawn planter. Hand-spraying equipment used on the smaller farm will still be adequate although the somewhat larger farm operator may want a larger and more sophisticated sprayer.

Moving to a commercial farm size where tractor power is available, the plow, disc and other implements for land preparation can be laid aside and a sprayer substituted for them. The only other change will be a no-tillage planter, one designed to cut through surface mulch and hard soil. Generally, tractor size is determined by the power needed to pull the planter, not the plow.

On plantation-size operations, equipment may include airplanes and special spraying units for rapidly covering the territory. Otherwise, the equipment is not necessarily different from that used on medium-sized farms. On farms of every size some cleanup work with small sprayers will have to be done in order to control difficult weeds. Therefore, the hand sprayer is a required tool on any size farm.

It is difficult to judge which farm size will adopt no-tillage most easily. From an educational point of view, the large farms should make the change with least effort. Their change will be somewhat constrained by the equipment already on hand and by limitations of the labor force. The group which may make the change fastest is the moderate-sized commercial farms. These farms are owned by people who are aware of machinery, agricultural chemicals, varieties and fertilizers. They are able and willing to adjust to methods which can make them more money.

The poorest group of farmers probably will change slowly. However, within this group there are innovative farmers who can be expected to demonstrate hitherto unknown advantages for no-tillage. Since this group makes up the largest number of people, it is imperative that special efforts be made to accelerate their acceptance of no-tillage if progress is to be made.

Soil Conservation

Advantages of no-tillage on any size farm are the ability to plant crops without the time or energy needed for land preparation, and to preserve water and soil from loss. The first advantage is one that recurs yearly. The advantage of saving water can be crucial some years and of no importance in others. Soil loss exerts a continuous and largely irreversible drain on soil productivity. At first, this effect is hardly seen, but eventually there are whole areas where no productive soil exists. Such places are easy to see in many parts of the tropics. Whether their significance is appreciated by the natives is another matter. It seems that, at present, there is a growing awareness of the need for soil conservation throughout the world.

In the tropics, rains of very heavy intensity are the rule. In the natural situation, the soil is covered with trees, brush and grass so that little soil erosion occurs. With clean cultivation the soil is exposed so that it is highly subject to erosion. Typically the soil can be removed to the depth of loose soil (5-10 cm) during a few rains. Thereafter, erosion slows but the larger gullies grow ever larger. In contrast, soil that is covered by a mulch of dead plant parts is resistant to erosion, almost as resistant as soil in pasture.

During the past several years, the authors have had the opportunity to see soil erosion problems in Latin America and in Southeast Asia. In some mountainous areas as much as a meter of soil has been lost in 20 years after the land was opened to cultivation. Deforestation – pasture-cropland-pasture-brushland – is the sort of haphazard rotation practiced on much of this land. Declining fertility and lower water-holding capacity are the principal factors which reduce soil productivity so markedly in a few years.

Because of population pressures and economics there is no way for tropical countries to return major portions of the land to forests or to permanent improved grassland. It must be faced that the land is going to be used to grow food crops regardless of its quality and regardless of its slopes. The question that needs to be asked is how can this be achieved with the least soil degradation. Thus far, the best answer is the use of no-tillage or reduced tillage where crop residues are left on the soil surface.

For this no-tillage revolution to be achieved will require a vast effort of research, extension, demonstration and, perhaps, coercive legislation. Probably none of this will be forthcoming. The pace of change that will occur in the tropics will no doubt be slow and great losses of soil will occur. There probably is no way to avoid this waste. In general, people do not respond idealistically to perceived problems. Instead, near total disaster has to occur before anything at all is done. Soil erosion is such a disaster in the tropics. It is now being seen as a disaster, but it is likely that progress towards a conservation tillage solution will be both spotty and painfully slow.

On the optimistic side, there is hardly a country in the world where some work on methods of reducing tillage is not being done, however fitfully. There currently is easy interchange of scientific results between countries and rapid availability of agricultural chemicals. It is not impossible to hope that, given this start, no-tillage will yet prove to be an immense force in the tropical countries of the earth.

REFERENCES

Aubert, G. and Tavernier. 1972. Soil survey. In *Soils of the Humid Tropics.* National Academy of Science, Washington, D.C., pp. 17–44.

Baver, L. D. 1972. Physical Properties of Soils. In *Soils of the Humid Tropics.* National Science Academy, Washington, D.C., pp. 50–62.

Buol, S. W., F. D. Hole, and R. D. McCracken. 1973. *Soil Genesis and Classification.* The Iowa State University Press, Ames, 360 pp.

de Andrade, V. A. 1982. No-till rice in Rio Grande Do Sul, Brasil. Peliotas, Brazil *EMBRAPA, No-Till Notes, Vol. 2, No. 1.* Office for International Programs for Agriculture, University of Kentucky, Lexington.

Drosdoff, M. 1972. Cultural systems on tropical soils. In *Soils of the Tropics, Prairie View – Texas A&M College Bull., No. 2,* pp. 13–28.

Goswami, P. C. 1971. Shifting cultivation in the hills of northeastern India. *India Farming* **21**:1–10.

Greenland, D. J. 1975. Bringing the Green Revolution to the shifting cultivator. *Sci.* **190**:841–844.

Hayward, D. M., T. L. Wiles, and G. A. Watson. 1980. Progress in the development of no-tillage systems for maize and soya beans in the tropics. *Outlook on Agric.* **10**:255–261.

Huntington, T. G., G. W. Thomas, R. L. Blevins, and A. Perez. 1982. The influence of tillage-systems and weed control practices on the production of corn and soybeans in the Dominican Republic. *Agronomy Abstracts, Am. Soc. Agron. Meeting,* Anaheim, California.

Juo, A. S. R. and R. Lal. 1977. The effect of fallow and continuous cultivation on the chemical and physical properties of an Alfisol in western Nigeria. *Plant and Soil* **47**:567–584.

Kalpagé, F. S. C. P. 1974. *Tropical Soils.* Macmillan of India Limited, New Delhi, 283 pp.

Kendrew, W. G. 1961. *Climates of The Continents,* 5th Edition, Oxford University Press.

Lal, R. 1975. Role of mulching techniques in tropical soil and water management. *ITTA Tech. Bull. 1,* Ibadan, Nigeria, p. 38.

Lal, R. 1981. No-tillage farming in the tropics. In *No-Tillage Research: Research Reports and Reviews.* Editors, R. E. Phillips, G. W. Thomas and R. L. Blevins, University of Kentucky, pp. 103–151.
Reports and Reviews. University of Kentucky, pp. 103–151.

Lal, R. 1976. No-tillage effects on soil properties under different crops in Western Nigeria. *Soil Sci. Soc. Am. J.* **40**:762–768.

Lal, R., D. De Vleeschauwer, and R. M. Nganje. 1980. Changes in properties of a newly cleared tropical Alfisol as affected by mulching. *Soil Sci. Soc. Am. J.* **44**:827–833.

Lal, R. and D. De Vleeschauwer. 1982. Influence of tillage methods and fertilizer application on chemical properties of worm castings in a tropical soil. *Soil and Tillage Res.* **2**:37–52.

Manipura, W. B. 1972. Influence of mulch and cover crops on surface runoff and soil erosion on the land during the early growth of replanted tea. *Tea Bull.* **43**:95–102.

Marelli, H., B. M. de Mir, and A. Lattanzi. 1981. La temperatura del suelo y su velacion con los sistemas de labranza. *INTA. Informe especial 14.*

Mensah-Bonsu and H. B. Obeng. 1979. Effects of cultural practices on soil erosion and maize production in the semi-deciduous rainforest and forest-savanna transitional zones of Ghana. In *Soil Physical Properties and Crop Production in the Tropics.* R. Lal and D. J. Greenland, Editors. John Wiley and Sons, pp. 509–519.

Meyer, R. J. K. 1975. Temperature effects on ammonification and nitrification in a tropical soil. *Soil Biol. Biochem.* **7**:83–86.

Muzilli, O. 1981. Manejo da fertilidade do solo. In *Fundacao Instituto Agronomico do Paraná 1981.* Plantio directo no estado do Paraná, Circular IAPAR 23, 244 pp., pp. 43–53.

Muzilli, O. and K. Ique. 1979. Evaluation of tillage systems and crop rotations in the state of Paraná. *Instituto Agronomico do Paraná (IAPAR) Progress Rep. 1976–79.*

Olson, R. A. and O. P. Englestadt. 1972. Soil phosphorus and sulfur. In *Soils of the Humid Tropics.* National Academy of Science, Washington, D.C., pp. 82–101.

Perez, Aridio. Personal communications. 1982. CENDA, Santiago, Dominican Republic.

Roman, E. S. and M. R. Barker. 1978. *Anais do II Encontro Nacional de Pasquisa sobre Conservacao do solo, Passo Fundo,* 347 pp.

Sanchez, P. A. 1976. *Properties and Management of Soils in the Tropics.* John Wiley and Sons, New York, 618 pp.

Shenk, M. 1979. Unpublished data. CATIE, Turialba, Costa Rica.

Soil Survey Staff. 1975. *Soil Taxonomy, Agric. Handbook No. 436,* USDA. U.S. Government Printing Office. Washington, D.C.

U.S. President's Science Advisory Report. 1967. *Tropical Soils and Climate.* U.S. Government Printing Office. Washington, D.C., pp. 472–500.

Wijewardene, Ray. 1980. Energy-conserving farming systems for the humid tropics. *Agricultural Mechanization in Asia,* Spring, pp. 47–53.

Zaffaroni, E., H. A. Burity, E. Locatelli, and M. Shenk. 1979. Influence del no laboreo en la Producción de maiz y frijol, en Turrialba, Costa Rica. *CATIE, Report FITO 872–79.*

Index